Integrating Modern Science with Recovered Ancient Science ...

Control Centers of the Body
and their Interface

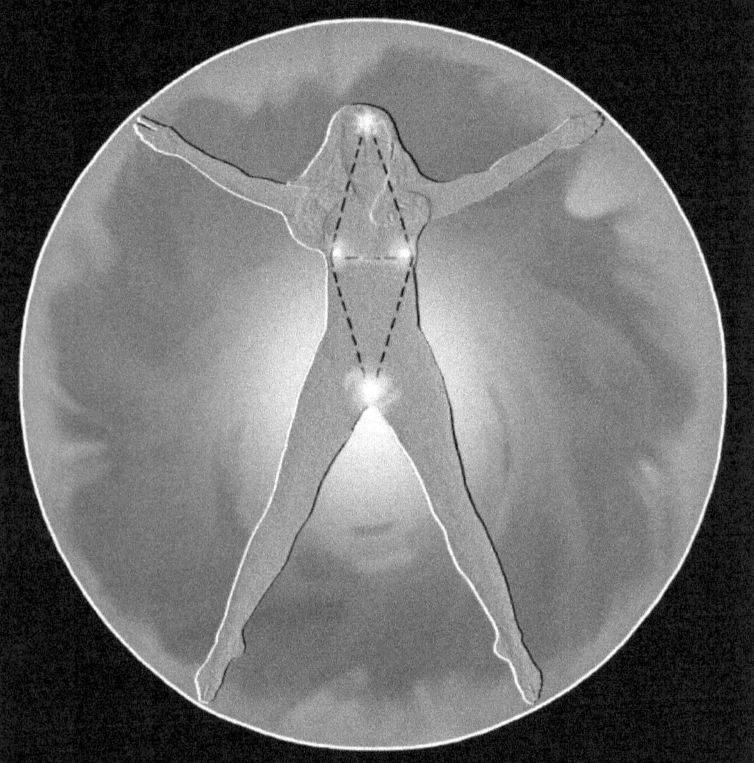

Head - ***mental*** **control center**

Breasts - ***thought*** **and** ***feeling*** **interface**

Heart - ***physical*** **control center**

Two Major Forces of the Body and their Symbols

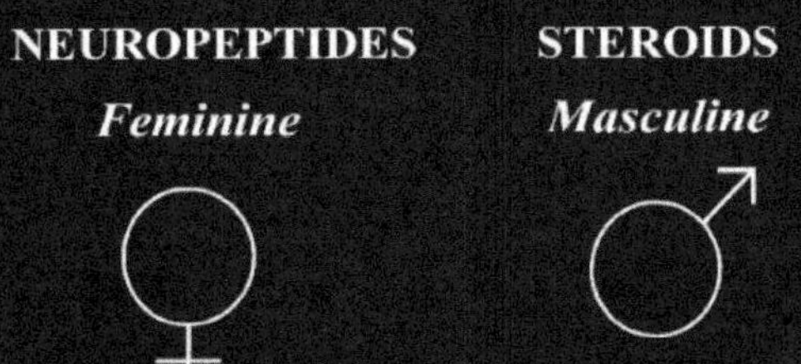

Four Basic Elements of the Physical World

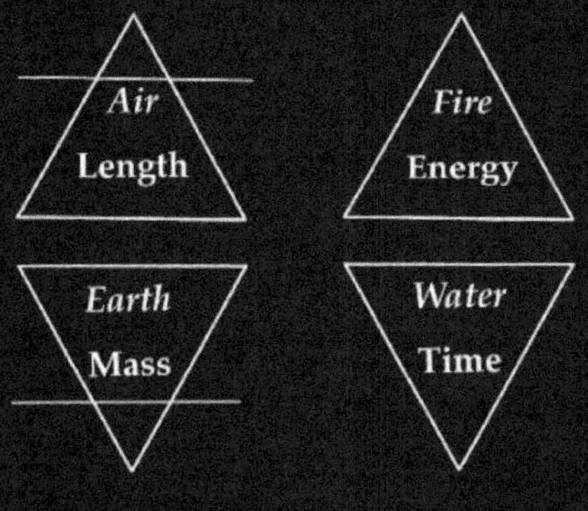

Integration

Controlling Your Hormones

Power From Above, Within, and Without

by

Robert L. Peck with
Leslie M. Cassinari and
Christine S. Gavlick

Integrating Modern Science
with Recovered Ancient Science

Lebanon, Connecticut
U.S.A. 2008

www.personaldevcenter.com

2 3 4 5 6 7 8 9 10

About The Cover

The cover suggests the history of controlling the inner forces of the body and mind. The chalice symbolizes the inner transformation caused by the mind as it activates a new persona. The Greek *liknon* depicts the swelling within the physical control center of the body which initiates the secretion of a controlling fluid or hormones. The drugs are the modern external replacement of the inner secreted hormones as the source of transformational change.

Publisher's Note

This is a second printing, January 2020, of the original 2008 publication. It contains minor editorial corrections throughout the text as well as format improvements in the Indian Sanskrit translated documents (i.e. *Haṭhapradīpikā*, *RigVeda*, and *Rudrayāmala*). In addition, in the frontispiece illustrating the "Four Basic Elements of the Physical World," the Earth and Water triangles were drawn incorrectly in the 2008 publication and have been corrected.

Cover design and graphic art by Leslie M. Cassinari
Translations of parts of *Haṭhapradīpikā*, *RigVeda*, and *Rudrayāmala* by Robert L. Peck

Personal Development Center
Post Office Box 93
South Windham, CT 06266-0093

ISBN Paperback: 0-917828-12-7
ISBN Paperback: 978-0-917828-12-6

Library of Congress Control Number: 2008930225

Printed in the United States of America

Table of Contents

Controlling Your Hormones

Power From Above, Within, and Without

Introduction

Controlling Your Hormones was researched and written to explain one of the oldest, shortest, and yet most comprehensive teachings of the ancient world, namely the *Rudrayāmala* of India. The *Rudrayāmala* has been highly condemned because of its description of the inner powers of individuals that can be self-controlled without external political, religious or social forces.

This book leads the reader through historical evidence of the truths described in the *Rudrayāmala* up to their present day confirmation with modern endocrinology. It then concludes with the translation and commentaries of the complete *Rudrayāmala*. But let us begin near the beginning, over two thousand years ago.

The early *Mithraic* soldiers can be viewed as one of the first known groups to have turned the search for higher powers into a science. Obviously desirous to survive on the battlefield, these soldiers were highly motivated to develop greater strength, speed, anticipation, courage, and a working oneness with their comrades. Since soldiers travelled widely and encountered the knowledge of many different peoples, they assimilated a great deal and naturally tested out their new knowledge on the battlefield. Their findings provided a foundation which evolved into the prominent *Dionysian* movement in ancient Greece and Rome. Their control of inner powers was described by their detractors as resulting from secret underground rituals during which an intoxicating and transformational libation known as *haoma* was served from a chalice to the participants.[1] Little more is known about the *Mithraic* rituals, since positive descriptions of them as well as the *Dionysian* rituals were either not written down or more likely destroyed.

Fortunately, insights into the *Mithraic's* secret knowledge of *haoma* can be gleaned from surviving writings and artifacts from other ancient sources. This is because of the universal sharing of philosophy, religion, and science between cultures before the Common Era. For example, the Persian god *Mithra* is found in India as *Mitra* and was certainly known in other cultures as a solar god.[2] The transformational libation *haoma* can be equated with *amrita* in the Sanskrit writings, and *amrita* can be related back to the Greek word *ambrosia* which also had the meaning of being a transformational inner elixir.

Modern research into the ancient world reveals universal common themes in the ancient writings and artwork which serve as clues about the inner powers. For instance, the digging out of Pompeii uncovered extensive wall paintings depicting the *Dionysians* as did the exploration of the hidden Catacombs of Rome. Three common themes related to higher powers were the chalice, the libation, and the inner source.

The Chalice

The concept of a chalice was widely used in ancient times to denote a container for a metaphysical libation commonly interpreted to have heavenly powers. However, in order to make sense of many early writings and

[1] Peck et al. (2006), Ch. 5

[2] For example, the Roman god *Sol Invictus*

artifacts as well as modern endocrinology, the concept of a chalice must be extended to the body and even to specific organs of the body.

As an introduction into the concept of the body being a chalice, consider the ancient Egyptian alchemical *Athanor*. The *Athanor* was described as the container in which alchemical reactions take place to produce gold out of lead. The size was described as being similar to that of the human body. There is another well-known example in the New Testament story of Jesus converting water into wine for a small party, but the water was described as being within rain storage vessels (*hudrion*, υδριον), each of the size of the body. Similarly, it was a universal practice to describe metaphysical powers as residing within an individual's body as if it were a chalice.

It is of interest to note that the public generally views the chalice itself to have more of a transformational power than its inner content. This is because the actual power of the transformational elixir is generally highly dependent upon ritual which includes not only the mode of libation but also the expectation. It is always a chalice which is raised in a toast or promise — whether the actual container is an ordinary tumbler, a golden goblet, or a gourd. It is always the dedication associated with a chalice which contains more power than whatever the contents or actual container might be. The power of a physician or healer was related far more to the presentation of medicine than the actual medicine, and it is almost certain that more Placebo Effect cures have been obtained with a sugar pill or bitter tasting tonic than with all of the antibiotics ever routinely prescribed.

Similarly, there are references to organs within the body as chalices, such as the inner loving heart which becomes the source of transformational powers. The ancient science was able to describe this process in detail with the heart making the toast or promise at the central table of the body.

The Libation

The mystical chalice used in rituals to impose special powers contained an elixir which the Greeks characterized as heady, intoxicating, and able to carry an individual upwards.[3] The transformational powers were attributed to a spirit, power, or god that was present in the elixir but which was nonetheless subject to those sharing the libation. The inner power, for instance, could open or close minds but its action depended upon those partaking of the libation.

The Greeks considered that the libation of wine from a properly dedicated chalice could initiate the opening of doors to the higher realms of the mind of each participant. Intoxication from the elixir consisted of the liberation of the mind from social bondage and restricted thought and was a very positive and necessary step in evolving. In Sanskrit the word for intoxication was *madya* or the "state of being mad" with the same positive meaning.

[3] Greek: ανωρής *anores*, "to carry upwards"

The allegorical god *Dionysus* became the liberating and intoxicating power which allowed an individual to break free of conditioned responses and thoughts and open the body and mind to what had been placed in the heart or required to meet a need. To find the power of *Dionysus*, a participant had to first be in the radiance of *Apollo* or in the state of dedication, openness, and expectation. The power of *Dionysus* then managed to diminish or obliterate the conditioned binding self-image, thoughts, judgments, guilt, and self-importance. It was this release from societal control that led the rulers and authorities attempting to dominate and control society to describe *Dionysus* as advocating intoxication, erotic behavior, and madness.

Perhaps there is no better spokesperson for the power of libation than Plato. In *Symposium*[4] only a select group of people were invited, definitive rules were set, and a subject to be explored was agreed upon. The amount of wine consumed was just the amount that could release the participants from their limiting self-images and concerns. In *Phaedrus*, Plato explained that the power of *Dionysus* could then open the united group to the *Muses* who became the source of the feelings and insights for the chosen subject. The individual powers of *Aphrodite* and *Eros* then interpreted the input of the *Muses* into a language acceptable to everyone.

The power of the libation was to open the mind to inspiration from the heavens and to find a union with others who shared in the libation and the subject of search. Needless to say, many people experience this with or without drinking an intoxicant and often with the shar-

---

[4] Greek: "drinking together"

ing of food. The sharing of food and other physical pleasures became other methods of bringing forth the power of *Dionysus*, but they all required the initial dedication along with trust and openness to one another.

The Inner Source

As the ancients evolved their methods for the transformation of the body, mind, and personal world, no doubt, they came to the obvious conclusion that the source of transformational power was a directly controllable fluid within the body rather than a spirit in heaven because of the correlation between their practices and the results on the battlefield.

They looked for an inner chalice or source of the mystical fluid, which was found to be in the center of the perineum so named since in Greek *peri* means "all round" and *neuma* means "control." The inner chalice symbolized the swelling of the area of the *bulbospongiosus* muscle — a bulbous, porous, and expansive muscle in the middle of the center of control. This inner chalice had the power to respond almost instantaneously, which the other chalices were unable to do.

The swollen bulb in the perineum was symbolized in Greece as the *liknon* carried in parades by the *Dionysians*, which not only portrayed the swollen muscle but also the method of its stimulation. The *liknon* contained a cloth-covered bulb in the middle of a winnowing basket. The size and shape of the winnowing basket depicts the bottom of the body and the cloth covering depicts the skin covering the swollen gland of the perineum. The winnowing basket futher indicated the winnowing motion that takes place in the lower guts during the swelling and the release of the libation.

Instead of depicting a masculine perineum, the Indians portrayed the swollen bulb as rising out of the female pudendum and described its excitation like either the churning of butter or the winnowing of grain.

Chapter One
Hormones and the Ancient Nectar of the Gods

Ancient science can be assumed to have begun with the development of weapons along with studies of how to use them. History books describe the teaching of the manual of arms and various battle formations and tactics, but they ignore the studies about increasing the inner power of the individual soldier to move and respond faster. I did considerable reading between the lines when I was researching what was known of the highly successful *Mithraic* soldiers and their secret underground practices. I jumped to the conclusion that they had learned how to deliberately increase their steroids. They could then be compared with modern athletes who have researched how to increase their powers with pharmaceutical steroids.

Only recently has the concept that an individual can increase personal powers through a consumed chemical become even a possibility. Prior to the knowledge of the power of hormones, the power and skills of athletes and soldiers were considered to be based primarily upon genes and training as well as what was often called luck or divine intervention. Modern science, however, has not yet explored how individuals can generate and control their own hormones, but rather has concentrated on changing their inner powers with pharmaceutical drugs.[5]

---

[5] Cooper et al. (2003)

The ancient science that correlates with modern endocrinology[6] is still ignored and even suppressed because the ancient science has been labeled as pagan and pantheistic. This is because both the sources and effects of the inner powers were described with allegorical gods, while the hormones were described as inner mystical nectar and not as biochemicals. The ancient and modern sciences dealing with inner powers can be initially assessed by comparing their basic descriptions of the inner controls of an individual, which, upon close examination, make the same statement.

Modern

Reproduction and behavior in a human is largely controlled by two classes of hormones, steroids and neuropeptides.[7]

Ancient

Reproduction and behavior in a human is largely controlled by two forms of the power of Love expressed in Greece with the allegorical gods, *Eros* and *Aphrodite*.[8]

Eros and the steroids are described with nearly the same characteristics, the chief being of course, sexual drive. *Aphrodite* and the neuropeptides are described as being the source of socially interactive forces, and the nonapeptide ("nine peptides") oxytocin is rapidly gaining a reputation as the loving, trusting and cuddling hormone, which is the early concept of *Aphrodite*.

---

[6] Endocrinology includes the study of hormones.

[7] Adkins-Regan (2005)

[8] Lucretius (1993) and Plotinus (1991)

It was not until modern endocrinology started describing the powers of hormones in anthropomorphic terms that the two sciences could be directly compared. Both sciences differ greatly from the physically oriented, classical physics view of the world in their acceptance of metaphysical powers within individuals. Both the ancient science and its modern counterpart agree that the inner hormones can be controlled to varying degrees both from within and without. The ancient science suffered from the lack of present day knowledge of physiology and directed its energies to studying the inner control of hormones and their effects. Modern endocrinology suffers from the loss of the ancient methods of inner control and instead studies how to control the inner powers of the human body through externally administered drugs.

In the modern West, the only popular and readily available references to the ancient awakening of the inner powers are in the Bible where soldiers are told that they should prepare for battle by girding[9] on their swords and armor and then girding their loins with the obvious intention or need of obtaining increased speed, strength, awareness, and creativity. There are no extant Western writings as to how to gird the loins, yet many people have the inherent feeling that it is related to the tightening of the groin experienced when stepping into a cold shower or facing some physical threat. Fortunately, India did retain records of how to produce inner hormones[10] by girding the loins with a practice called *bandha* which has the same meaning as "girding."

[9] Hebrew: *azar*, "to compass, to bind about"
[10] Called by various names described later

Ancient Indian science also studied and described the gain in individual powers with *bandha*.

Proving the Ancient Claims

It seemed obvious to me that to prove if there were any truth in the ancient claims, the first step would be to determine if the inner girding or *bandha* could produce increased outer physical powers that would be useful to a soldier. Accordingly, I set up a controlled experiment comparing how fast individuals could physically respond to a sound. I had three volunteer groups. The first was a group of adult yoga students who had been taught how to gird, the second was an untrained group of adults, and the third was a group of children.

The three groups were tested by measuring the time it took them to respond to a sudden, unexpected sound by raising their arms and clapping their hands. The response time of the adult students was then compared with that of equal-sized groups of "ungirded" adults as well as 12–13 year-olds who would normally be expected to respond much faster than either group of adults. The result was startling, since the average "girded" adult responded 2 times faster than the average ungirded adult and 1.4 times faster than the average ungirded youngster.[11]

I was well aware that the girding might not have produced modern warriors and that the subjects' response times might be slower than that of a trained boxer or fencer, but it did prove that girding could do at least part of what was claimed by the ancients. That proof

---

[11] Appendix p.189 or Peck (1998), Appendix A.

also allowed me to accept the detailed writings of Sanskrit as being based upon truth and worth the effort to extract their teachings. It also increased my interest in describing the ancient claims in terms of modern science. The Greek root for the word hormone provided an excellent start, since it comes from *orme* which means "an impulse or urge." I related this meaning to the words *ambrosia* in Greek and *amrita* in Sanskrit, both of which meant "to enliven" and were used to describe the inner nectar that "urged" the inner gods to do the bidding of our hearts.

I certainly liked the description given by the Greek philosopher Plato and other philosophers that it is through Love that an individual gains control of his or her world. This Love was not the same as desire, but rather in the nature of a goal or purpose in life which was administered by the two inner gods of Love, *Eros* and *Aphrodite*, who can be directly compared to the two main divisions of hormones, steroids and neuropeptides.

Technology and our modern industrial world was built on centralized outer controls, but I believe that in creating an externally uniform society and world, we have gone too far in ignoring the inner source of power within individuals. Certainly, there is rising social unrest and a breakdown of bonding between individuals[12] which is, after all, the root of society. Modern society controls us by teaching constant judgment of self and others which overloads the body with the stress hormones, instead of encouraging the acceptance and trust

[12] Putnam (2000)

of our own friends which increases the neuropeptides such as oxytocin.

Suppressing the Inner Powers

If the ancient method of girding the loins to awaken the inner powers of individuals is as effective as it appears, why did the knowledge of it disappear off the face of the earth? The early *Mithraic* soldiers[13] kept the nature of *haoma* highly secret, which makes good sense since they surely did not want their future opponents to learn of it and use it against them on the battlefield. Plato, who certainly knew about *ambrosia*, stated that rulers would not want their subjects to develop powers as great as their own and hence he predicted that future rulers would suppress information about increasing inner powers.[14] This can be construed that rulers did not want their subjects to know of the power of Love, and so knowledge and control of the heart and breasts (of both men and women) became the first organs of the body to be suppressed.

How this suppression took place is worthy of a book of its own because of the variety and effectiveness of the methods used, but in brief, sometime after the fifth century the ancient model of the inner powers of the body was drastically changed. (Chapter Three gives a detailed view of the old model.) The nature of the heart in the sacred (*sacral*) region and the emotional reactions of the breasts were declared to be in the beating heart in the chest. The sacral and perineal regions of the body were then declared to be filthy and needing to be kept under tight tension to avoid leaks. Men's breasts were

---

[13] Ulansey (1989)

[14] Plato *Symposium*, paragraph 64.

defined as mistakes or the remnants of fetal development. Maturing children were taught that the breasts of young girls were extremely vulnerable to injury and must be protected at all times and never touched. (This is certainly not true and is now known to be detrimental for the development of the nipples.)

The nipples and breasts of boys were embarrassing in general as they were considered to be feminine and not at all masculine. Hence, they were ignored and seldom touched to insure that they wouldn't grow or become more feminine. Children's gut feelings were ridiculed and particularly so since the feelings were often in opposition to imposed societal beliefs. The sensual feelings in the breasts were stated to be non-existent, and instead, feelings were stated to originate from the beating heart as were the yearnings in the guts.

The ancient Greek word *stethos*, for the two breasts of either gender, is defined in the Greek-English Lexicon as, "the seat of feeling and thought as we use heart."[15] In other words, the ancients were quite logical and scientific in ascribing feelings to the sensitive breasts instead of to the beating heart, since the heart is not known to have any sensitivity to Love other than to the physical excitement or stress of the body.

The obscuration of the ancient universal views of the body and its science was ensured by the burning of related Western writings that described the original terms and their functions. Extensive Eastern documents, such as the Sanskrit documents of India, were not burnt but suppressed just as effectively by creative academic

---

[15] *stethos*: στηθος

translations undertaken around 1900 that were acknowledged to discourage the ancient science and religion.[16] There was the persisting belief that only royalty or the religiously inspired could have, at best, a limited access to any kind of higher or metaphysical powers. The commoners were just that, common, and when any ordinary person did become a hero, saint, or genius, it was explained as being directly controlled by heaven for the purpose of heaven.

That the suppression is actually continuing in full force is evidenced by the denial of the ancient universal model of the body and the lack of acceptance of the old descriptive anatomical names. Instead the body and brain are being described largely in mechanical or computer terminology by scientists (with the exception of a growing number of endocrinologists). The effects of shifting the location of the heart is well evidenced in the general population by the atrophy of the perineum, loss of vitality, reliance upon external drugs, the large market for pads for incontinence, as well as the speedy decay of the brain and body with age.[17]

A Modern Example of Inner Powers

Perhaps the best example and method I have to introduce the ancient science and its concepts is to use the personal story of the Nobel Laureate Otto Loewi who exemplified the reliance upon inner powers and whose resulting research brought a starting basis for accepting the scientific reality of the ancient nectar of the gods. Loewi was strongly motivated to research and discover

---

[16] *Rigveda Samhita*, Vol. 1, Ch. 4, pp.103-106.

[17] Peck et al. (2006) Ch. 13. Also Resnick, et al. (2003) on Kegel in *JAMA*, July 16, 2003

how the organs of the body could communicate together because it could not be explained by the existing knowledge of his time.

Loewi related how he half-wakened from a dream[18] with the feeling that he understood how organs communicated[19] and wrote down a sketchy summary. The next morning after fully awakening, however, he had the very frustrating experience that his conscious brain was incapable of understanding the summary. Nevertheless, Loewi managed the next night to completely ignore his thoughts and instead opened to another vision. His dedication was rewarded with another dream, but this time before he could lose the message, he dashed into his laboratory to manifest the weird image of his dream.

The experiment which he managed to contrive consisted of placing two beating frog hearts into two separate solutions and then noting the rate decrease in the faster beating heart when some of the solution from the slower beating heart's container was added to the faster heart's container. The result was extremely important to science since it equated a control and intelligence to an inert solution and dissolved chemical.

Reviewers of his work are amazed at how he overcame many of the obstacles in making his experiment function (particularly at 3 a.m.) His final results, however, gave positive evidence that the organs of the body could be stimulated and controlled by an inner secreted biochemical instead of by only nerve or electrical im-

[18] Diamond (2006)
[19] Donnerer & Lembeck (2006)

pulses as previously commonly accepted. His experimental results also allowed each organ or center of the body to be considered as a separate life form, controlled by hormones generated under a central metaphysical control in the sacral heart, a basic concept of the early ancient science.

The Author's Fortunate Experience

In order to introduce my role in connecting Loewi's work with the ancient science, it is first important to note that the majority of scientists do not believe Loewi's story of getting knowledge through a dream nor do they believe the dream of the Noble Laureate Friedrich Kekule about a snake biting the end of its own tail which took organic chemistry to a new height. Scientists tend to state that it is only the outputs of the conscious brain, following lots of applied methodology, hard work and study, which produce progress in science. Certainly not lying in bed sleeping and dreaming!

I, however, found that many scientists are privately open to metaphysical sources of knowledge. One day at the lab while I was on a coffee break, the conversation turned to Transcendental Meditation, then becoming popular. I stated that I had found something even more potent for kicking up my creative juices, but that it could not be discussed in polite society.

Imagine my surprise when word of that statement got around and one by one many of the staff "sneaked" into my office for a private disclosure. To get some work done, I finally had to call a meeting in the auditorium and describe a few of my experiences. This, however, led to further demands until I was teaching evening classes through adult education programs. Since I was donating my time, I felt that it was only fair that the students become guinea pigs or subjects in experiments investigating various ancient self-development practices, which they agreed to.

We did finally verify and modify the technique I had been using, which amounted to a relatively simple exercise that certainly reduced mental tensions and at the same time increased vitality and awareness. This exercise started with sitting in a chair, quieting the mind, and rocking slightly on a rolled cloth placed under the perineum. The breath was constricted to become a labored exhalation pressing down to the perineum while reaching for very pleasant feelings and listening for a ringing (tinnitus) centered between both ears.

I sent one of our resulting text books[20] to the well-known Aurobindo Ashram in India asking for comments. This induced the head of the Ashram to visit and spend two days with me as he became interested in what we were doing. He wrote that we were working with the basics of "*Tantra*, *Veda*, everything."[21] After visiting, he encouraged me to read the original Sanskrit writings and wrote an article later about his experiencing the universal nature of knowledge or *gnosis* in America.[22]

I was again surprised when I finally started to read the ancient Indian *RigVeda* in its original Sanskrit and

---

[20] Peck (1976)

[21] M.P. Pandit (1977) No. 54

[22] Ibid., No. 56

found that the exercise developed for my classes was clearly outlined in the 28th chapter of the first book. (See Chapter Twelve.) It stated that to develop higher powers in the body one *begins* with quieting the mind, pressing the perineum, finding a good feeling, rocking and exhaling strongly in order to press out a juice (the juice called *soma* had been described earlier as *amrita*, meaning "nectar of vitalization"). It was at this time that I suddenly became convinced that *soma* and hormones were one and the same. The *RigVeda* text describes that the perineum swells, the abdominal muscles churn, and fluids flow down as well as up. I then got the definite message that I was just beginning.

Shortly after that, I focused my efforts on translating other Sanskrit documents and studying related modern scientific documents. I had to ignore the existing Sanskrit translations since they were quite at variance from the literal Sanskrit.[23] I found support in my research from reading nearly identical condemnations of "mystical" groups such as the Gnostics, Mithraics, Dionysians, Eleusinian, Alchemists, Tantriks, and Arians, etc. Their detractors agreed that whatever these groups were doing was powerful (and threatening) and that it involved deviant and licentious sexual practices. There was also common agreement that the powers were related to some mystical fluid which was generally falsely assumed to be alcoholic or hallucinogenic.[24]

I could now make an educated guess that the early groups were first stimulating the body through lower body motions and stimulating the mind with visions. In

[23] See the comparison of translations in Ch. 12.

[24] Peck et al. (2006), Ch. 16 pp. 104-105 & pp. 117-118.

turn, this would cause the production of hormones which would have included dopamine, endorphins, serotonin, prolactin, histamine, and cortisol. These are known today to have very powerful controls on the body and mind as well as oxytocin known to increase union and bonding between people.

The strong and united defamation of these mystical groups and their members certainly indicated that they were conceived of as presenting a threat, but not a physical or violent threat to the underlying power of ruling institutions. The threat had to be that they were perceived as presenting evidence that rulers did not have a special Divine Right and covered their weakness with robes of state and their ignorance in laws.

A Personal Testimony

I feel that I should conclude this chapter with my own experience, since I am eighty years old with a more active brain and body than I have had since childhood. I caution those of you who want to reverse the effects of aging that it may take months to finally awaken the old responses of the body, but I can also offer encouragement that there is no end to the awakening process. My wife and I are constantly stating that our present world together far surpasses any earlier time in our lives and that we have not yet found any leveling-off in the ever-increasing ecstasy, union, vitality, sexuality, and creativity within our world. With this awakening, we have also found the truth of the ancient statements that there is no death, only an unending opening to more.

Therefore, please accept that the following chapters are not only based upon careful scientific and literary methodology but also upon personal experience of their truths. However, they are by no means complete and each reader must add to what is presented herein with his or her own experiences.

Chapter Two
Metaphysical Cause and Physical Effect

Perhaps one of the hardest concepts for a hard-working, dedicated, and well-versed scientist to accept is that major breakthroughs in science can come from dreams, visions and mental games rather than years of hard physical work and study.

Imagine the irony of the story of Otto Loewi going to bed for his second night to find the answer of how hormones communicate. In our materialistic society, it is a rule that gain must come with physical pain and not with dreaming. The accepted cause for the betterment of the world today is in rallies, conferences, mass demonstrations, and mailings describing what must be changed. In science and education, for example, greater funding and increased political controls allegedly cause changes.

What are the available causes of happiness, creativity, and increased strength other than the highly questionable taking of club drugs such as ecstacy? How acceptable is the universal, ancient dictum that the cause of obtaining whatever is desired in life can only be found within the individual breaking free of social bondage or conditioning? This teaching has little meaning since the modern materialistic world cannot accept the concept of metaphysical causes, much less their control.

The loss of the acceptance and understanding of metaphysical causes was due to the rise of classical physics. Ironically, classical physics owes its power to the very surprising fact that every physical object, force and interaction in the universe is manifested by only four metaphysical causes. These universal metaphysical building blocks were known three millennia ago as *earth*, *air*, *fire* and *water*. Over time the four metaphysical causes became more sophisticated. The Earthy element became identified with *mass*. The spatial nature of Air became defined as the property of *length*. Fire became identified with the effects of *energy* including motion, temperature, electrical current, and light intensity. Water as the element describing fluidity or physical changes became *time*.

The modern elements of *mass*, *length*, *energy* and *time* are just as metaphysical as the original elements and cannot be defined or measured; however, they each have a dominant physical effect which can be catalogued and standardized. For instance, *mass* is now related to a standard weight, *length* to the length of a standard bar, *energy* to a repeatable source of energy, and *time*, of course, to the movements of a clock. But it should be emphasized that no one knows what causes *mass*, *length*, *energy*, or *time* or what they are. They remain metaphysical although their effects are physical.

The present situation might be compared to a bakery which uses four basic ingredients — flour, sugar, eggs, and milk — to make many different products that are designated by their many different forms, sizes, tastes, and textures. The final products seem to have very little correspondence to the properties of the ingredients in much the same manner as physical matter and its interactions seem unrelated to their four building elements.

In order to illustrate the mystery of metaphysical elements and their causes, let me relate a personal story. I had developed a simple device containing a solution and two inserted metal plates which produced electrical power when heat was passed through them.[25] The results were important enough to gain me patents, financial backing, and finally a meeting with some top scientists in the energy field to explain how the device worked. The device was extremely simple, and despite my addition of special ingredients into the solution, the only explanation that was required was how does the flow of heat manage to change into electrical power able to drive a physical motor?

It is amazing how the group fell into the implication of this not so simple question. It quickly raised two questions, What does heat have in common with electricity? and How can one be turned into the other? Despite all of the training and experience of the members of the group, this simple childlike question brought the group down to a childlike union and appreciation of the deep metaphysical forces of nature. We even considered our own personal musings such as, How does a bar of lead know that it is heavy? or How does light go through glass and not metal?

Science answers many such questions by describing the cause as a "field." It is the gravitational field which causes things to have weight, it is the molecular field within glass which causes light to pass through. I finally came to the realization that the physical sciences have made progress because they have learned how to

---

[25] *Thermogalvanic Cells*, US Patents 4,376,155; 4,410,605

ignore metaphysical causes. This observation is supported in the requirements for obtaining a patent. The U.S. Patent Office, in evaluating an invention, does not require an inventor to be able to explain how or why an invention works. The Patent Office will not issue a patent on a cause, only on a demonstrable effect.

The ancients considered cause (and its field) of equal importance to its effect and that to fully understand something, both had to be known since they were both necessary to bring forth something from chaos. The causative force was defined as being masculine, as often symbolized by the sun, while the manifested effect was defined as being feminine, or of the moon. The sun as a primary source can be understood from the Greek word *henotheism*[26] used during the time of the worship of *Sol Invictus*. *Henotheism* can be interpreted to refer to:

1) the metaphysical cause or source (sun),
2) the field (radiation) and
3) the effect (heating) as being of one nature.

The alchemical *Emerald Tablet*[27] is the early Egyptian expression of the same truth. It speaks of how an object or action consists of both the masculine and feminine, and if the world is to be correctly observed, the physical must be separated from the metaphysical.

Modern science, in strong contrast to ancient science, describes Laws which state what must always occur

---

[26] Greek: *heno*, "one in union" and *theism*, "from the gods." Also, see Peck, et al. (2004) pp. 7 & 228.

[27] *The Emerald Tablet – Tabula Smaragdina*. Also, *The Scaffold for Perfection*, Appendix.

under well specified conditions but which do not explain what causes the required results. Science must of course give some explanation for what caused the effects, but this is typically done by using metonymies which operate by stating that the cause is the same as the effect. For example, the definition of energy, *energeia,* is given by the inner (*en*) work (*ergon*) that is released, which is of course an effect not a cause. Heat is caused by fire, weight is caused by gravity,[28] darkness is caused by lack of light, starvation by lack of food, etc. I felt that I was progressing when I discovered that our society is maintained by a similar lack of separating cause from effect, or not separating the metaphysical from the physical. This ignorance of cause is due to our proclivity to substitute belief for cause without subjecting the effect to scientific proof or quite often even personal experience.

Art as a Cause

The early Greeks made a science of Art, recognizing it as a causative agent. The usage of the Arts to convey a cause was called *techne* (*tecknikos*)[29] and the various sources of specific causes were described as *Muses*. The Frontispiece is an example of *techne* which may be able to radiate the content of the entire book once the symbols are understood.

One of the best current examples of the nature and power of *techne* is the Placebo Effect. This power behind the ability of a sugar pill to give the same physical effects as an active drug can be allegorized as *Mnemosyne*, the mother of the *Muses*. (See Chapter Six.)

[28] Latin: *gravitās*, "heaviness"

[29] Peck et al. (2006) Ch. 4

The subject remembers what a drug is supposed to do and then utilizes that view to cause the hormones within his body to match the view. Unfortunately, no one appears to be interested in spending millions to study *Muses* and almost everyone seems to prefer taking a drug.

I was searching for information about when and how *techne* and the *Muses* became lost, when I obtained a book by Walter Lowrie who gave an answer to my questions. His book, *Monuments of the Early Church*, gave very interesting observations and analysis of the early artwork in the Catacombs. He describes the majority of it as existing without any ecclesiastical content[30] and that it was a continuation of Greek art or *techne*. He noted that most visitors to the Catacombs were impressed with the atmosphere or "field" of peace and the assurance of eternal interactive Love radiated by simple, nonprofessional artwork.

The earliest Catacomb artwork had three dominant metaphysical themes: the soul, interaction, and union. The soul was allegorized by the goddess of love, *Aphrodite*. Interaction was allegorized by the love of the Good Shepherd (similar to *Psalms* 23) as well as the Greek god *Eros*. Union and vital energy were symbolized by the fish, no doubt, taken from the ancient *Vesica Pisces* symbol for the united region of two circles or different worlds.[31] The usage of two fishes symbolized the astrological sign of *Pisces* which de-

---

[30] No suffering, crucifixes or dogma. See Parsons (1896) "Symbols".

[31] See Frontispiece inside the hexagram, union of male and female.

picts the resurgence of cosmic energy from chaos as well as the union of *Aphrodite* and *Eros*. *Pisces*, therefore, had a number of comforting meanings for mourners in the Catacombs.

The sense of union in the Catacombs is further reinforced by the drawings of small gatherings of people observing the traditional *agape* or the feast of love and union. There is apparently little doubt that these underground intimate love feasts were the continuation of the *Mithraic* underground love feasts with the object of finding union through the sharing of the stimulation of *ambrosia* or *haoma* which could be described today as the individual generation of the hormone oxytocin. Grieving mourners, caught up with the emotional pain of the immediate physical death of a loved one (physical effect), could view a painting imbued with *techne* and touch the spirit of its message (metaphysical cause) and thus be transported to timeless peace, immortality and union with a mutual love.

Lowrie notes a striking change in the artwork of the Catacombs created around the fifth century. He attributes these changes to the rise of the power of the Church and its insistence that all artwork be ecclesiaastical or supportive of centralized teachings or laws. This required definitive art rather than any works open to personal interpretation. Lowrie noted that, as a result, all artwork began to have written titles or descriptions for added verification of proper meaning.

Today, television advertisements are probably the best examples of how far this shift from the early Greek Art has taken place. Even our poetry and literature has become quite specific with very little left to the imagination and writing is judged by its clarity rather than emotion or feeling. Institutions hire communication experts to assure that only one meaning can be obtained from any publication.

Soft Science

There was, at one time, an effort for all sciences to follow classical physics by using its standard metrology methodology. However, modern life sciences had difficulties in using the four metaphysical elements as a basis for physical measurements. As a consequence of not having universal metrology standards, the life sciences have been categorized as "soft" sciences despite their attempts to eliminate any subjective data and references to metaphysical causes.

It is easy to understand that under this measurement criteria along with academic censorship, ancient writings were denigrated as being superstitious and pantheistic and completely outside of science.

The dream of Otto Loewi can now be conceived of as a threat to the very foundation of endocrinology as a science and even to classical physics when he attributed intelligence and control to inert nonliving chemicals. Science was facing the scientific proof of causes other than the four fundamental elements. Life had to become accepted as being in violation of the Laws of physics.

Before Loewi, life had been found able to reverse the normal erosion and decay or entropy of matter as each life form was able to evolve under its own inner control and resist the well established "running down" of the universe. Life was more identified with the Creator of the universe (metaphysical cause) than with the Body (physical effect.) After Lowie, life could be more identified with the physical processes of the body. His experiment proved the power of the body to control itself.[32]

I am convinced that we are living in a very exciting age of major changes which is perhaps easily evidenced with the title of an article I just read, *Serotonin Modulates Behavioral Reactions to Unfairness*.[33] Endocrinologists are now stating in a formal scientific paper that a chemical in the body can alter how an individual reacts to observing other individuals being treated. This article is only one of hundreds equating hormones to causing changes in the relationships and feelings of individuals to their outer world.

There seems to be an ever-increasing belief in the inner powers within individuals that supports the ancient acceptance of the possibility of an individual becoming as a god. To fully understand the ancient science, it is necessary to take the next step of describing and defining some of the causes that can affect an individual.

---

[32] See Ch. 4, pp. 40-41.

[33] Crockett, et al. (2008)

Chapter Three
The Ancient Model of the Inner Self

A simplified early Greek model of an evolved individual can be introduced by the following diagram:

Chaos ↔ ♀♂ *Zeus* ↔ ♂ *Psuche* ↔ ♀ *Aphrodite* ↔ ♂ *Eros* ↔ ♀ *Soma* [34]

Zeus is the source of creation of the world, *Psuche* is the "inner-most-me," *Aphrodite* is the "feeling-prudent-me," *Eros* is the "controlling-me" and *Soma* is the "physical-me." The symbols ♀ and ♂ stand for the feminine and masculine natures. The arrow indicates an interface without merger but with transfer of energy and intelligence.

Chaos ↔ Zeus

Chaos was defined to be the ultimate source of everything which is or will be manifested including *Zeus*. *Zeus* is the name for the original source of identity, existence, creativity or evolution. *Zeus* as Creator is assumed to have both masculine and feminine natures. *Zeus* is the interface between *Psuche* and *Chaos* and is the source of Good, morality, *henosis* (union) and *gnosis* (knowledge).

The Inner Me (*Psuche*)[35]

The ancient model of the individual started with what we know today as the innermost "*Me.*" This is the *Me* who upon awakening suddenly in the morning asks

---

[34] Hesiod, *Theogony*; Aristotle, *Metaphysics*

[35] Sanskrit: *ātman, brahmā*

without any fear or concern, "Where am I?" and "Who am I?" It is the same *Me* who I and many others have experienced during what is called the near-death experience (NDE). When the NDE was first widely reported in the national news media, most people were highly interested in the fairly common descriptions of a tunnel, light, intense peace, pleasure, and the sense of a universal Love. However, in reviewing my own NDE, I also was impressed that I knew without a doubt that I was dead, but that knowledge caused absolutely no concern or fear. My brain was dead or at least inactive and certainly not involved; otherwise, I am sure it would have been quite hysterical with the realization that I was dead.

Normally, people are unaware of the existence of their innermost *Me* because they are overpowered by their brains conditioning as to what they are supposed to be. Most individuals have been conditioned to ignore the inner *Me* and listen only to the chatter and judgment of the brain. Hence, it becomes possible to identify the innermost *Me* with the death of the chattering brain during a NDE. One of the chief problems is that most people identify the *Me* with conditioned thoughts and therefore deny the true *Me*.

Other experiences such as the absence of reality upon awakening are assumed to be products of the brain and not a separate state of being. This assumption is also supported by the fact that the brain in its imaginations can produce thoughts similar to the NDE. The inner *Me* can, therefore, only be fully evidenced with its interface with the future or that which is not available to the physical brain. It is the hero's story of foreseeing a route through danger or the experience of finding an inner answer to an impossible question. The dreams of Loewi's *Me* (but not the consequences) as described in Chapter One were certainly denied by most scientists. This leads to another type of problem and that is that the inner *Me* and the brain cannot communicate together at all well with words but rather communicate chiefly by means of feelings.

One question I had was, Where does the inner *Me* reside, if it becomes more noticeable with the death of the brain? In reviewing ancient writings as well as the experiences above, I firmly accept the ancient description that the innermost *Me* resides in the heart in the bowels which the ancients described as the seat of the inner *Me*. Furthermore, it is this region of the body that may die long after the brain allowing the NDE before the entire body dies. Many cultures consider that the soul rises from the lower body such as depicted in the early Roman Catacombs.[36] The perineum is similarly described as the root and dwelling place of the inner *Me* in many ancient descriptions such as Indian and Chinese.

I feel a little foolish in using the word *Me*, but I know of no English or other modern word to describe it accurately without other possible meanings. However, the ancient Greek word, *psuche*[37] (ψυχη: "psoo-kay"), gives an excellent description of the inner *Me.* The *Psuche* was defined as the eternal nature of the self that was immaterial, moral, intellectual and creative as well as the center of existence, which I could immediately

---

[36] See Lowrie (1923).

[37] Liddell and Scott (1891)

fully identify with. It was this power of *Psuche* that was relied upon by many people and the basis for the observation of many ancient philosophers that enlightened individuals, who follow their *Psuche*, can do no wrong and need no teaching.[38] It was also the power of the *Psuche* which was credited by the ancients as being the root for the attaining of union with others and the world. To the ancients, it was the *Psuche* which gave individuals the powers to act like gods.

Over time, the meaning of *Psuche* changed in response to the fear of it by rulers. The Modern Greek Dictionary in *Strong's Concordance* offers an excellent example of what happened over the millennia. It states that the modern word *psuche* was derived from the word *psucho* (ψυχω: "psoo kō") which has a physical meaning of being the inner vitality of a life form evidenced by breathing. Modern Greek, therefore, describes *psuche* as being primarily the source of vitality and life rather than as an immaterial, eternal inner *Me*. This modern interpretation is, however, unsupported by ancient Greek writings.[39]

The Feeling and Prudent Me (*Aphrodite,* Αφροδιτη)

The concept of knowing Truth, Good, the future, or union with others is largely absent in modern English, although the word "prudent" has a residual meaning close to it. The word "prudent" is derived from the word "providence"[40] which has been ignored because of its metaphysical nature.

---

[38] Matthew 9:12-13, Goble (1970) pp. 28-29, and Ch. 11–13

[39] Liddel and Scott (1891)

[40] This has a different root from prude, which comes from prudery.

The first step in becoming as a god begins quite logically with the knowledge that you have special prudential powers. For instance, this moment is generally expressed in terms similar to the following:

"I had the feeling that I knew what was going to happen, and I found my body getting all prepared."

"We were able to feel and respond as one!"

"It took me a long time to explain my sudden feeling of the future."

"I suddenly stopped as I felt danger, and I consequently saved my life."

Almost everyone is aware of having had a sudden instantaneous feeling that included a complete awareness of the surrounding world that was far more instructive and important than those received from the normal five sense organs.[41] This feeling can include empathy, sympathy,[42] intuition, apprehension, *gnosis*, compassion, sentience, and commiseration. Feelings are the tools which unite the self and the world into a state of oneness or union such as found in dreams.

The organs for perceiving feelings as well as the comprehension of them has already been introduced in Chapter One with the ancient Greek word *stethos* which meant the breasts of both sexes and the seat of feeling and thought. Before continuing, it can be helpful to understand that the early meaning of *stethos* has

---

[41] See *Techne*, Ch. 6. p.43.

[42] Greek: συμπάθειά: *sumpathaeia*, "fellow-feeling," "sympathy;" πάθημά, *pathema*, "that which befalls one."

been considerably altered in Modern Greek[43] which argues that the word now means the entire chest, bosom or heart (beating.) The effect of this modern distortion can be illustrated with the atrophy of aging breasts with their loss of the accuracy and power of feelings that are replaced with imagination or conditioned judgments.

With a bit of reflection or experiment, it is possible to verify the role of the breasts in interacting with the outer world. Consider a time when you broke down and cried deeply. Crying must start with observing or imagining some tragedy. This may begin with a strong exhalation or inhalation if the exposure to the tragedy is sudden, unexpected and intense. As will be described later, the breath can initiate the release of steroids to activate the entire body to physically respond if required.

In response to tragedy, the breath changes to a slow, laborious exhalation with the sense of an increasing tightness or fullness around or in the breasts. It is at this time that one can describe this as feeling the rise of the urge to cry. There is the rising sense of constriction in the breasts and difficulty to either deeply exhale or inhale. It is as if, there was a tight band placed just below and above the breasts. If observed carefully, the constriction is like that obtained during the girding of the loins. The breath will generally continue with labored short breaths. In the case of horror, you can report the feeling of being unable to breathe at all.

It seems reasonable to assume that the constriction in the breasts is due to the forced increase in blood flow

---

[43] Strong (1990)

similar to that of a blush. This increased blood flow can be surmised to increase the sensitivity of the breasts which further increases the feelings of the tragedy or source of sorrow.

It is at this time that a feedback between the tragedy and the reaction to the tragedy occurs. As the breasts become energized, the feelings associated with the tragedy also increase, and as the feelings increase, the sense of constriction increases. This feedback is sensed as the emotions seem to keep increasing with each labored short breath. It is also during this time that the eyes can well up with a great deal of tearing which serve to reduce the excess steroids[44] and assist in the rebalance of the hormones.

It is the instantaneous sharing of feelings with the increased activity of the breasts which has been called the language of Love. The breasts as sensors and thought organs are able to interface between situations and the interactive individuals such that thoughts, feelings, intentions, and needs are communicated without barriers of language or even distance.

Because of the importance of this language of Love, the ancient Greeks developed an excellent allegory to define the power and nature of *stethos* with the anthropomorphic goddess *Aphrodite.*[45] The name itself helps define the allegory, since *Aphrodite* means "created from foam" or the interface between sea and land or symbolically between the metaphysical and the physical. It seems obvious that her *stethos* were her sensors

---

[44] See Frey & Langseth (1985).

[45] Sanskrit: *Devī*

and had to be described as being prominent and responsive as Homer did nearly three millennia ago.

In attempting to find a model for *Aphrodite* with her prominent breasts and communication powers, my thoughts sped to the cartoons that depict tough middle school teachers as being big-breasted. I personally knew a teacher who seemed to have extra senses and seemed specially endowed with empathic powers with her awareness of boys like me who might try to make a denigrating gesture behind her back or lie about having read the assignment. Needless to say, little boys like me were scared silly of big-breasted teachers, and this fear and respect were, no doubt, felt by the original creators of *Aphrodite*.

Over time, the world stopped relying upon the feelings and thoughts of the breasts, and all members of society had to rely upon what they were told to feel by institutions and written laws. Anything connected with individual feelings and thoughts had to be suppressed. That this suppression is still continuing is evidenced in the replacement of the feelings of competent teachers with mandated curriculums and school-wide discipline policies capable of being executed by every teacher. Similarly, the feelings and reactions of mothers in addressing their own children's problems are being replaced with standardized and approved behavior models.

Over the millennia, any references to nipples and breasts other than for nursing or sexual eroticism have become nonexistent. One obvious example of this lack of recognition of their importance is the undeveloped nipples of expectant mothers and the small size of men's nipples. Society's rulers and institutions can gain control over individuals by suppressing and denying the power of feelings and instead enforcing the power of obedience.

Without an active *Aphrodite* and *Psuche*, it becomes impossible to be able to discern truth from accepted statements and platitudes or from imagination and desires. Two very common examples are the inability to discriminate your own health from hypochondria and the blind acceptance of false statements. This lack is often stated as the confusion felt with a strong gut reaction and its opposition to some social standard or rule. Without an active *Aphrodite*, any interface between the outer world and inner *Me* is extremely difficult.

For more information about the sensory powers of the breasts refer to Chapter Nine which attempts to provide basic physiological data as well as historical references.

The Controlling Me (*Eros*, Ερος)[46]

The center of the controlling *Me* is in the perineal or groin region which exists in the center of the outstretched body (See Frontispiece *Control Centers of the Body*.) The power of this center was described as being that of the allegorical god *Eros* because of its individualistic and anthropomorphic properties as will be described. Being in the central location, it optimizes the speedy control of motor functions of the body necessary for the "spontaneous" response to pain and preservation of the body as well as for programmed phys-

---

[46] Sanskrit: *Iśvara, deva*

ical responses such as used by dancers and athletes. The early soldiers engaged in hand-to-hand fighting were perhaps the first to become aware of its importance and its interconnection to the breasts or to the "instant feelings" of combat. A foot soldier had to react and had no time to think.

The controlling center or *Eros* also has the power of activating the body to conform to an expected feeling such as that of becoming a king in a child's game, becoming a hero on the battlefield, or becoming a wise advisor to a friend. A person "putting on" a professional demeanor uses and relies upon *Eros* to effectuate the change and to furnish the needed added powers.

Aphrodite or the feelings must communicate with *Eros* or the control system with a special language which is nearly instantaneous with the capability of conveying prudence. This language was described as the special language of *Psuche* which was able to convey the description of an entire world in an instant. *Aphrodite* is able to use the same language to integrate the subject of the language into the present and future and then describe it to *Eros* to manifest it. The language can be described as conveying that which is yearned for or Love. *Aphrodite* and *Eros* therefore became the allegorical goddess and god of Love using and responding to the language of Love.

Modern endocrinology has no recourse except to assign to hormones the power to respond to and convey the language of Love. As for instance, oxytocin carries with it the message of union and its fulfillment, while the adrenal hormones carry the strong message of self-protection and how it can be maintained. This statement reminds me immediately of the claimed powers of the inner elixirs such as *ambrosia* of the Greeks and *soma* of the Indians.

The source of intelligence for the interaction of *Aphrodite* and *Eros* was ascribed to the sacrum at the base of the spine.[47] The sacrum or sacred bone contains gray matter identical to brain cells. The sacrum was considered to hold the stored dedication or goals of the heart and the direction that the individual had chosen to follow. Functionally the sacrum becomes as a second brain guiding the autonomic motor response system of the body which is known to react even before the brain is able to know something has happened.

Eros has the power to completely turn off and reset the current thoughts of the brain and actions of the individual. This power is absolutely necessary if the body is to respond to an emergency or sudden demand. The "reset" power is obvious. When you are threatened, your mind is often described as going blank, while *Aphrodite* or your breasts respond and define the feelings related to the threat and then send its "thoughts" to the perineum and the sacrum. The sacrum then uses what is stored in its memory and then defines the general direction to be taken: fight, flight, placate, hide, protect, seek further, etc. The sacrum/perineum in many cases is able to communicate through the central nervous system, but other times it must marshal the hormones of the body to change the nature of the entire body such as during an extreme physical threat or mental demand.

---

[47] See Stross (2007).

Modern science is unclear about the mechanism of releasing hormones. The ancient Sanskrit description of releasing *soma* (as hormones) will be described using modern terms since it is well supported by experience.

Aphrodite directs the awakening of the perineum which promptly tightens (if not already tense such as during stress) and initiates what is well described as churning or winnowing. (That state of a clenching anus and the guts getting in an immediate uproar is a common method of describing this initial stage.) The motion of the perineal muscles is coupled to the abdominal floor and together they lift the entire lower abdomen up and down in a motion suggestive of a churn. This motion along with neural signals stimulates the adrenal glands as well as the intestines to produce the activating and energizing hormones. This activation was described quite well in a *Scientific American* article[48] about the adrenal hormones preparing the body for a challenge as:

> *"... epinephrine is the one handing out guns; glucocorticoids [49]are the ones drawing up blueprints for new aircraft carriers ..."*

The activation of the perineum with the churning of the lower body that resets the brain can be extremely pleasurable. This explains why children like to tell spooky stories, play wild games, perform athletic feats and, of course, directly stimulate the breasts and perineum. It can also explain those crazy people who love to jump out of planes or hang by their fingernails on a cliff or perhaps even that quiet, refined woman who goes shop-

[48] Sapolsky (2003)

[49] A group of corticosteroids from the adrenals.

ping for an expensive, outlandish outfit. It can also explain the power in a good cry, a strenuous workout, a thrilling movie, or a tough mental challenge.

A problem arises when the adrenal hormones vitalize the body with insufficient directions from the heart to cause churning or further mobilization of the body. With the added energy of the released corticosteroids, the brain can regain control and interpret the outer world through the sense organs and its own exaggerated conditioning rather than through the *Aphrodite*.

The control of *Eros*, therefore, shifts from *Aphrodite* and the Love stored in the heart to the thinking, conditioned brain. Since there was only a momentary stoppage of the brain and no resetting, the individual continues on under increased adrenal hormones and increased mental control which results in what is called the modern stress syndrome. With the direct control of *Eros*, the desires, fears, imaginations, etc. of the brain are amplified. As Plato explains,

> *"Eros directed to self is the underlying cause of all the offenses done by an individual; for the lover is blinded by the beloved, so that he judges wrongly ... and thinks that he always ought to prefer himself to the truth."*[50]

The Physical Me (Soma, Σωμα)[51]

As mentioned before, the *Psuche* is difficult to experience apart from the physical body, but it is not difficult to separate the body from the *Psuche* or from its effects. The Greeks described the purely physical body with or

[50] *Republic*, Bk. 9, paragraph 60
[51] Sanskrit: *jīva*

without a *Psuche* as *Soma* (not to be confused with *soma* in Sanskrit related to hormones.) There are two expressions which express two opposing relationships of *Soma* and *Psuche*, "The spirit is willing but the flesh is weak," and "His greed overrode his dedication." It was common in the early philosophies to describe *Psuche* as a masculine force and *Soma* as a feminine force. *Psuche* was compared to the sun, evolutionary and controlling, while *Soma* was of the moon and earth, both nurturing as well as nurtured.

The *Soma* or physical *Me* of humans, similar to that of animals, exists in a state of immediacy or being controlled by the perceived immediate physical demands of the world and the body. This immediacy response of individuals was cultivated by societies and their rulers to maintain a complex, stable and productive mass of people.

In order to control, society trains individuals to believe the conditioned social concepts as to what and who they are as well as what they are to think and do. One fundamental rule of society is that individuals must believe that they are powerless to change themselves and have to trust their outer world. This is, of course, quite in opposition to living in a world of your own worlds and identities without a central ruler or laws.

Chapter Four
The Nature of *Good* and *Evil*

The power of both *Good* (from *ghedh*[52]) and *Evil* which were well known to our ancestors are still with us today but their descriptions are hidden in our language. The usage of the term Love as causing miracles is a direct remnant of the old Greek gods of Love, *Aphrodite* and *Eros*. Plato stated that, "*Good* is anything which unites an individual or object with beauty (or perfection)."[53] Aristotle gave another definition, "*Good* is that for which all things aim."[54] The power of *Good* carries the vision of what is desired and controls the uniting of the pieces of the vision in order to manifest the vision.

Perhaps the best example of the metaphysical concept of *Good* is given quite surprisingly by science, rather than by the humanities. Almost everyone would have to agree that the greatest example of *Good* was the creation of the universe which manifested the present world. Science describes this evolution as beginning with the existence of basic elements or building blocks dispersed in chaos which also contained a special and concentrated energy. Science then assumes a step-by-step integration or union of larger and larger or more and more complex pieces out of the original building blocks. Each step was controlled by what science calls Natural[55] Law which includes a direction to be followed and the type and amount of energy required. The direction is variously described, but one re-

[52] The word "Good" comes from the root *ghedh* meaning "to unite, join or fit together as one." The word "god" comes from the root *gheu(ə)* meaning "to call, invoke."

[53] *Republic*

[54] *Nicomachean Ethics*

[55] "Natural" means metaphysical existence and dispersed in nature.

cent description is that the direction of evolution is always toward an economic or efficient result that has some derived advantage from that particular step.[56] Each step is, therefore, an example of the power of *Good* which, as Plotinus[57] explained in the third century CE, is always in the presence of an even higher level of *Good.*

In as much as the evolution of the universe is a fundamental example of *Good*, so likewise, the reversal of evolution is an excellent example of *Evil*. One large change occurred relatively recently in science when it accepted a metaphysical power it called entropy.[58] Entropy disunites and enervates the created world, driving things back into the chaos from which it came. Of course, science uses the well-defined terms of entropy and Law instead of the vague terms of evil and goodness. Entropy is now a scientific term normally used to quantify how much imperfection exists within something. The Greeks described entropy within individuals by how much they missed their goal in life (*hamartano*: originally "to miss the mark," now known as "sin"). Entropy would be comparable to a student's test score that indicates how much is not known or is lacking.

Once entropy was defined as a metaphysical quantity, classical physics could state that the physical world was indeed like a created machine that was now wearing out. Everything was knowable and everything had an end.

However, science also knew that there were evolving individual living entities which were increasing in size and complexity in complete defiance to the decay or entropy

---

[56] See Dunbar and Schultz (2007).

[57] See Plotinus, *The Enneads*, 1st Ennead, 4th Tractate.

[58] Greek: *entrope*: "turning toward" (chaos).

of the rest of the world. With entropy, science now had a powerful tool to perceive that life was an entirely different type of creation and was able to define it. Life could be perceived as manifesting a metaphysical power which could counter entropy and which was constantly becoming more complex with an increasing creative power of changing the inert and dying world.[59]

Life could be considered as a creative process that was at least as great as the original creation of the universe. Each living entity was, in fact, manifesting the power to unite energy and inert matter to form a reproducing, continuous, and evolving singular structure. Each living entity had the same potential power as the original Creator and could be as a god. Scientists could now understand the old philosophers who equated humans with gods (although they do not do so publicly.) Instead of viewing the Earth as dying, science began to see the Earth as the spawning ground of potentially evolving life forms containing inner creative forces or gods.

As science looked deeper into the evolving entities, each separate living entity was found to respond as if it contained its own intelligence and an awareness of its own reality and needs. Each entity also had an inner (but limited) power to change the self and surrounding world. This is evidenced quite well as each plant or tree vies for sunlight with extended growth. Life is characterized as having a network among inner cellular components, cells, organs and even other life forms. For instance, the very smallest and weakest of life forms such as the simple virus develops a network with other viruses to change themselves to counter antibiotics. Insects can unite to

[59] Schrödinger (1992)

form a colony that can share tasks to survive against forces that no single insect could do. Science discovered that there is a metaphysical force of *Good* that exists at the single living entity level as well as for a group.[60]

Classical science has been quite averse to explaining anything with metaphysical forces such as *Good* and *Evil* and has explained evolution of species in terms of measurable characteristics, such as survival and reproducetive rates which form the basis for the Theory of Natural Selection. However, due to increased knowledge, science cannot explain many oddities of evolution such as the rise in complexity of single cell life forms to multiple cells and then to multiple organs. This is particularly odd since the simpler forms continue to exist and then become the food for the more complex.

It should be noted that the ancient philosophers generally agreed that immediate evolution of individuals or groups is primarily due to their avoidance of pain and seeking of pleasure or the seeking of *Good*, which can, of course, result in what is called Natural Selection. This fundamental concept is appearing in advanced hormone studies wherein it is noted that hormones shift to increase pleasure and in turn affect the genes. This research may provide support for Jean Baptiste Lamarck's theory that evolution is in part influenced by experience.[61]

What is interesting is that it took the rediscovery of hormones as the inner transformational elixir of the gods to restore the ancient concepts of metaphysical forces such as *Good* and *Evil*. Science may just be now proving that universal metaphysical forces can indeed be experienced

---

[60] Vertosick (2002)

[61] Riddihough and Zaun (2010)

within individuals; otherwise, how could they be proven to exist anywhere? If they cannot be experienced within humans, how can we believe that we are anything other than robots?

Chapter Five
Metaphysical Forces

Almost everyone experiences metaphysical forces, yet, because of prevailing ignorance and widespread superstition, these forces are difficult to discern. *Aphrodite* is certainly a metaphysical force as is *Good*, but in preparing the reader for how these forces effectuate changes in the body, it is easier to start with a much simpler metaphysical force that controls the physical world, namely gravity. I use science as a sterling example of *Good*, and I also find that it offers excellent illustrations of other metaphysical forces.

Isaac Newton is the father of "hard," mechanistic, or Newtonian Physics that describes a purely physical world and is often believed to prove the nonexistence of metaphysical forces. Because of his materialistic physics, people are apt to be dismayed today when they discover that his chief passion was for metaphysics and that he believed in a Creator, all of which is perceived to be a contradiction to his physics.

Many people believe that he developed his theory of gravitation because he saw an apple fall. What he really saw was that the moon falls to the earth in the same manner as the apple, but the moon never reaches it. He promulgated the concept of an unseen and metaphysical force between the moon and earth that was the same as between the apple and earth. His critics strongly protested his characterization of gravity as metaphysical and claimed that it must have some physical explanation (which continues today.)

Instead of attempting to explain how the Earth attracted the moon or apple, Newton concentrated on the physical interactions and was able to mathematically equate the gravitational pull to the mass of objects, which equals mass times a constant dependent upon how fast an object would accelerate in the same location.

This equation describes a gravitational pull which can be compared to the feeling of a downward force in an upper accelerating elevator which is greater for a heavy person as well as for someone in a faster elevator. But because of Newton's explanation, the majority of people and scientists stopped searching for what caused the gravitational force in the first place. Gravity can be explained as that which attracts objects together, or gravity becomes defined *only* by its physical and measurable effects.

Newton's brilliancy lay in discovering that knowing the why or how of physical reactions was not necessary to explain what happened. In other words, physicists need not be concerned with causes but only the physical effects or pragmatism.[62] This led to modern science which is giving us marvelous creations, from space vehicles to computers. His hard science was so perfect and complete that most of the world, including the majority of scientists, have now come so far as to believe that metaphysical forces do not exist.

However, times are changing and now many scientists, exemplified by Einstein and Loewi, are proving that the metaphysical forces upon which science was resting can no longer be ignored. The four sacred fundamental units of measurement: *mass*, *length*, *time*, and *energy*, which

---

[62] Peirce (2000)

had long been assumed to be purely physical, now have to be considered to have much the same prime nature as the ancient and universal metaphysical elements of *earth*, *air*, *water* and *fire* from which they were derived.

There is increasing evidence that the modern elements of measurement are not even immutable as once believed. Quantum, Particle and Astro Physics are demonstrating how the basic nature of particles and fields can be altered with energy exchanges. Time was found to be relative and might even turn around and go back into the past.

With modern instrumentation, the soft sciences such as biology, physiology, medicine, and psychology are now proving the existence of metaphysical forces that, although reported in ancient literature, have been largely denied in the modern world. Consider two well substantiated studies. The first is a very exciting medical discovery called the Placebo Effect which is demonstrating an inner metaphysical force of humans to change their inner hormones to mimic a drug. The second is the work done by Princeton University's PEAR group[63] which has given solid evidence of the metaphysical power of human concentration to slightly change the functioning of precision machines[64] which is, no doubt, the inner power to mimic a drug.

In order to develop a method of describing these forces and others to be mentioned, it is necessary to build upon both the methods being used by modern physics as well as the methods used by the ancients. Physics invented the term "field" to name that which connects metaphysical sources to their effects as well as to other sources and ef-

---

[63] See Ch. 7, also Jahn and Dunne (1989).
[64] Jahn, R., & Dunne, B. (1986)

fects. For instance, the weight of an object exists because it is in the field of gravitation. The strange phenomenon of a moving magnet inducing an electric current to flow in a nearby wire is explained by the existence of an electromagnetic field.

Similarly, Princeton's PEAR group studied the field of consciousness which exerts a small but quite perceptible force on external objects as will be discussed in Chapter Seven. The usage of the term "field" can certainly be viewed as a similitude to the ancient term "aura" associated with a god. The terms "field" and "aura" are descriptions of the physical or observable effects of an unexplainable inner or associated metaphysical force. The ancients would state that the aura of the sun causes growth of plants, whereas physics would state that plant growth is due to the radiation field of the sun.

The source of the aura was of more importance to the ancients than the aura, since they considered that metaphysical sources could be changed or controlled, which until recently, the modern world could not believe. However, as physics and endocrinology are continually bringing forth more and more data to support these ancient concepts, the next step is to look into the ancient views of the inner controlling energy and how it was directed.

Chapter Six
The Inner Transcendent Energies

The ancient philosophers correctly assumed that there were many different types of energies or powers within the body. The different inner energies which rule the body and mind can be compared with the different types of energy that power a city such as: chemical, electrical, mechanical, optical, computational, etc.

Similarly, the health of a city can be compared with the health of a body. If you are healthy, then the body and brain are able to meet the goals of your lower heart and evolve just as a city can maintain and expand its commerce.[65] Health, in other words, requires a supply of the necessary energy and effective control of that energy. If sufficient energy or control cannot be supplied, then illness will result for both an individual and a city as they both sink under the burden of unchanging rigidity, rules, and custom.

Plato described two forms of madness, one produced by illness and the other as a release from the yoke of custom and convention.[66] The latter release requires five powers allegorized by the gods: *Apollo*, *Dionysus*, the *Muses*, *Aphrodite* and *Eros*. These five gods can be compared with five words in ancient Sanskrit used to describe the same madness: *maituna*, *madya*, *maṃsa*, *mudra*, and *matsya*.[67] The two sets can be roughly compared as: *Apollo* and *maithuna* refer to the state of union or oneness;

---

[65] Schrödinger (1992)

[66] Plato's *Phaedrus*

[67] The Tantrik 5 "M"s. See Peck et al. (2006), Ch. 8.

Dionysus and *madya* mean madness, exhilaration or mania; *Aphrodite* is comparable to *mudra* which means your image, role, or physical soul; the *Muses* and *māṃsa*[68] refer to inner senses, feelings, and voluptuousness; *Eros* and *matsya*[69] are creative animation and manifesting vitality.

No existing writings of Plato describe how the lower nutritional bioenergy of the body is converted into the higher energies other than the statement that the *Muses*, the goddesses of the Arts, feed the soul and Gymnastics feed the body. Yet what are the *Muses* and Gymnastics?

Let me start by using the ancient Greek model of the *Muses* to explain the creative experience of Otto Loewi mentioned earlier. As he slept, his soul must have opened to the radiance of *Apollo* which filled his heart with a wave of new insights which initiated a hormonal response. The release of hormones awakened him and he jotted down only his feelings, but his notes were meaningless the next morning after he fully awakened.[70] The failure of this first "dream" can be explained as Loewi receiving the full content of truth without any of its modification to fit his limited memory or, in other words, the inability of his brain to understand "Apollo-speak."

He needed the mother of the *Muses* named after memory, *Mnemosune*, to shift the feelings of the dream to fit his own limited memory. Then he needed the daughters or the *Muses* to provide their ability to convert feelings and visions into understandable thoughts and images. It could

[68] As the flesh or inner succulence.

[69] *matsya* equals the Zodiac *Pisces* and the resurgence of cosmic energy from the depths.

[70] See Diamond (2006).

be as if the *Muses* were able to perceive *Apollo's* view of the hidden inner metaphysical and physical forces acting within the body and then use Loewi's current wisdom and experience to describe it. During the second dream, Loewi allowed himself to become maddened or overwhelmed with *Apollo* and the *Muses* and when fully awakened, he managed to dash off to his lab and manifest the feeling gained from *Apollo*.

Consider a personal experience of your own when you felt a strong need to explain to someone something that your brain could not begin to put into words. Generally, shortly after meeting this person you suddenly find words gushing out of your mouth which amaze you with their accuracy of expressing your feelings. In this case the ancient Greeks would have said that it was done by one of the *Muses*. Artists, musicians, actors, authors etc. are particularly responsive to the need to express some inner feeling to the outer world and this explains their dependency upon some *Muse* to interpret their feelings into various Art forms and why particular *Muses* have been named for their professions. The *Muses* were credited with using *techne*[71] to communicate between worlds and individuals.

The ancient science of *techne* is largely forgotten, but it was revived in part by the American philosopher Charles Peirce with his description of semiotics or the usage of signs to convey knowledge.[72] Peirce's semiotics and the Greek's *techne* require that if a sign is to convey a concept, there must be a relationship between the sign, the actual object, and its interpretation. Examples of signs are given in the objects on the cover of this book as well as

[71] Greek: *τεχνη*. See Peck et al., (2006), Ch. 4.
[72] *The Philosophy of Peirce*

in the Frontispiece with its depiction of the source of inner power.

The hypodermic syringe has a strong connection between the sign, the object and the interpretation. Very little needs to be said about what that sign means. Similarly, the ancient dwellers of the Indus Valley left the *Shiva Linga* as a sign for the people who followed after them as shown in the Illustration.[73] There is also a very strong relationship between the actual shape of the icon in the Illustration and its sign and its projected concept. Once the shape of the icon is recognized, the ancient meaning is quite clear but requires almost this entire book to explain it in words.

I also assume that much has been lost about the actual interpretation of the Gymnastics of the Greeks which were used to feed the inner transcendent energies of the body. Plato insisted that Gymnastics should not be concerned only with building muscles, but whatever he thought should be developed surely has been lost. A partial clue is obtained from the origin of the word Gymnastics which is from the word *gumnos* which means "nude, bereft or stripped". Gymnastics, beyond pure athletics, was probably a process of stripping yourself of something more than clothes. Practical experience teaches that physical activity can remove tensions, worries, and weariness and leave your body and mind open both to your inner self and to the outer world.

Young children at play certainly demonstrate how they integrate the stimulation of both the body and the brain. Many of their instinctual Gymnastics are no longer rec-

---

[73] See Illustration, The Muscles of the Perineum, p. 87.

ognized, such as sitting cross-legged on the floor. Perhaps one of the greatest Gymnastics, heartfelt sobbing, certainly changes the self and world. Unfortunately, the modern world interprets Gymnastics as belonging only to athletics or as developing muscles or a beautiful body instead of opening the body and mind to an awareness of and response to a higher state of existence.

There are a number of Indian writings[74] that describe the ultimate Gymnastics as churning the guts or moving them up and down to create a transcendent energy and using inner winnowing to purify or select that energy.[75] The winnowing is accomplished by assuming that the abdominal region is like a woven winnowing basket such as used to separate grain from its chaff. Winnowing requires both the rapid raising and lowering of the basket while letting the wind or breath blow away the chaff.

There are images in the remaining artwork of the early followers of *Dionysius* in Greece depicting parades with marchers carrying small winnowing baskets each containing a cloth-covered *lingam* (see cover) which can be assumed to correlate with Indian *lingams* as icons depicting the source of transcendent energy within individuals. Both types of *lingams* have been labeled as phallic symbols by detractors, but they do not have the right proportions, the glans, nor prepuce of a phallus.

In addition, another group of Greek artifacts that correlate with Indian writings are the paintings and statues that depict the Greek god of health, *Asclepius*. *Asclepius* is shown carrying a rod with a snake coiled around it, which is far from looking like an authoritative symbol as

---

[74] See Ch. 10–12.
[75] See Ch. 11–13.

is often assumed. The combination of the rod with the coiled snake is also depicted in Indian artwork where the rod is the symbol of the *lingam* and the snake is the symbol of vitality, renewal, and creativity.[76]

The inner rod as the *lingam* can be described as rising up from a fire pit[77] in the sexual region. The meaning of the name *Asclepius* provides a further correlation of the rod or *lingam* with both the Dionysians as well as the Indians, although the West now generally denies the name as having any meaning. *Asclepius* can be perceived as being formed from two separate word roots. The first is *askeo*, meaning "to be disciplined," while the second root, *leptos*, has the meaning of "being purified by winnowing."

The ancient models are certainly understandable if we consider experiences of being severely frightened. Inner churning starts with the clenching in the lower guts, the stomach rising up to the throat, and the sinking feeling in the tummy. The inner winnowing describes quite well the feelings many report with strong feelings of panic. The breath is fast and deep with strange sensations in the pit of the stomach as you seem to be reaching for understanding or, as the ancients might state, a purity of mind and body.

The modern world considers that the divine, enthusiastic and creative madness manifested by the five powers is a severe illness and generally labeled as manic depressive syndrome and treatable with sedatives and stimulants. It is often acknowledged that the geniuses of our society are highly creative during the manic phase but the public is

---

[76] *Sarpa*, the snake, was used as an allegory of continual renewal because it renews its own form or skin.

[77] *kundali* from *kunda*: "bowl of fire"

certainly conditioned to view the madness cited by Plato as requiring medication.

Certainly our schools suppress any rising mania in our children and even worse, individuals are made afraid of any sensations associated with mania or the loss of conditioned behavior. Is it not a severe oppression of a nation if the members are not taught of the existence of mania which not only can increase the joy and value of life but also the awareness of their own eternality?

Chapter Seven
The Science of Creative Love

Perhaps there is no greater concern to Western society than those individuals who seem to find some inner source of knowledge and power to control their own lives. Almost everyone — rulers, subjects and individuals alike — are horrified at the thought of someone not completely dominated by the rules and laws of society. In the modern world, laws are considered to be the cause and source of all that is good (or bad) even though a law has no power of its own and at best can only state what is supposed to be. This concept is maintained because good people are defined as following law, and hence the law must be good since it produced such good people. However, there are some individuals who perceive that their own *Good* or rightness is caused, not by external laws, but by some inner intelligence and power.

The modern world has considerable trouble in explaining the source of intelligence and power evidenced by either the Great or the Destructive people. As a consequence it is customary to refer to their powers as either god-like or satanic, with the source or cause of their actions and powers coming either from heaven above or hell below. The fact that both can be perceived as ignoring laws is quite troublesome, particularly since they seem to gain what they want without it. Some children learn this as they observe another child who breaks social laws and yet gains favors and attention, while they who adhere to the laws gain little. It is, no doubt, the conditioned belief that law can cause even greater pain or cause greater attainments of later rewards that keeps most children under control.

Since the Great and the Destructive people are not strict adherents to laws, they cannot be explained by laws and can only be compared to individual gods or demons who are each an allegory of some ultimate human power. From an academic standpoint, it is far easier to study and refer to a perfect model than a randomly distributed characteristic. Over the past millennia a great deal of effort has been spent in describing the heavenly or perfected powers; however, efforts in the last century or so have concentrated mainly on the hellish characteristics.[78]

The Greek poet and philosopher Hesiod was perhaps the first to attempt to catalog and describe all of the popular gods in his *Theogony*[79] which is generally assumed to be only a collection of myths. Plato, however, pointed out that Hesiod was using the created gods in *Theogony* to describe the nature of creative mortals rather than gods. Plato saw that Hesiod's *Theogony* is actually an allegory of the inner powers existing within each individual.[80] With this reasoning, other ancient writings such as those of Alchemy can likewise be viewed as an allegory of inner powers within individuals. The *Emerald Tablet* of Alchemy[81] says as much in its opening line, "As above, so below." The Alchemists describe inner powers as "immortal fire," *ignis innaturalis*, within an *Athanor*, which is difficult to perceive as anything but a human body.

Scientific academia cannot accept ancient descriptions of the creation of the universe as an allegory for personal creativity and presents it own theory of creation as fact.

---

[78] See Goble (1970), Ch. 1.

[79] Meaning "creation of gods"

[80] See *Symposium and Phaedrus.*

[81] Appendix

Nevertheless, the scientific model of the creation of the universe is nearly identical to the one that Hesiod gave over two millennia ago. Hesiod's short description was that the vision or Law of the creator god *Zeus* was manifested by the god *Eros* who selected the required energy and building blocks out of Chaos and formed them into the envisioned world. Compare this description with the modern scientific explanation that creation was caused by Natural Law which caused the combining of energy and building blocks released from chaos in an explosion to form the universe.

The only material difference is that Hesiod's description is more complete and describes a force, *Eros*, which is able to enforce or manifest the will of *Zeus*. The Natural Law of the Big Bang Theory includes within itself the enforcement and execution of the Law as well as the statement of what is supposed to be. This is obviously different from normal social laws which have no power of their own and require separate agencies to enforce or perform what the law requires. Hesiod's *Eros* can be defined as functioning as the agency enforcing or manifesting Natural Law.

These differences can be resolved by considering that both the ancient and modern descriptions of the creation require the interaction of four separate metaphysical powers to effectuate the creation:

1) the Vision or description of what is to be,
2) Chaos, the source of energy and raw materials,
3) a Directive Energy or *Eros*, which combines energy and the elements from Chaos in accordance with the desired vision, and

4) Time, the delay of evolution to match the forward rate at which reactions and the observations of them can take place.

Classical physics does not know how to enforce adherence to Natural Law but knows that somehow it is enforced. If this were not so, there would be no regularity in nature. However from a practical standpoint, since there is regularity, the metaphysical forces that enforce or manifest the Law do not need to be described. It is similar to how children respond to a law without questioning it, such as the law that they must take a nap.

Modern science has power since it can state what will happen without having to explain why or having to make it happen. To give an example of the current usage of Law by classical physics consider the Law, **W=mg**, that states in mathematical form that an object has a weight, **W**, proportional to its size or mass, **m**, in a gravity field, **g**, and that all objects must follow the Law. The Law can be perceived as an edict rather than an explanation of why a rock is heavy or what a gravitational field is or how it can move a rock. The Law only describes the final effect but not the causes, even though common usage treats the Law as a cause.

The nature of cause, *Eros,* or the power that makes things happen and change, was perhaps just as confusing to the ancients as it is today. In order to explain the allegorical god *Eros*, another allegory was made of *Eros* as Love. The power of that allegory continues to exist today and is often described as the powerful directive Love which exists in the lower heart of dedicated people. (Detractors of its power redefined *Eros* as sexual lust using the derived word “erotic”.) Poets have, however, stated such things as "Love makes the world go 'round," and that it can fulfill the dreams of the pure in heart. In his *Symposium,* Plato dedicated an interesting special session to the power of the god *Eros* or the god of Love. Plato began by asserting that there was little support for *Eros* but then brought forth many descriptions of its overwhelming power both for good as well as self destruction.[82]

The original Greek model of *Eros* as the allegorical god of Love and destruction was considerably weakened over the following millennia as *Eros* was denigrated to lust, and Love was limited to personal, sexual and romantic forces. However, science is now developing considerable support for the original concept of *Eros*.

Modern science knows far more about metaphysical energy than was known by the ancient science. Energy is now recognized as required for any change which occurs, whether moving in a car, cooking a meal, or thinking. In fact, the need for energy is even selling breakfast foods and candy bars. One large increase in the awareness of energy is that energy is not flowing free in the air but is locked up in its past forms by what science calls barrier potentials. One excellent example is the stability of a lump of coal which is quite inert until it is heated to release all of its contained hidden energy. The lump of coal can be compared to the mild mannered person walking home who is then ignited by noticing smoke coming from a neighbor's house and dashes in to save the trapped family.

There is a relatively recent discovery by science which has had a major influence on the concept of creation and

[82] *Republic*, Bk. 9, paragraph 60

creative energy. This discovery is that the physical universe, although still expanding, has stopped evolving and instead is wearing out and gradually disintegrating back into the elements of chaos.

This discovery required the recognition of the other power called entropy which somehow is able to counter the regularity and determinism imposed by Natural Law governing the Big Bang creation and subsequent evolution. This recognition gave rise to the *Second Law of Thermodynamics* that energy cannot be created or destroyed and that entropy can only increase or at best remain constant.[83] This Law has created great consternation which continues today in science verging on questions such as what constitutes reality. It can now be pointed out that Hesiod's creation story describing the creative directive power *Eros* did not have this problem, since *Eros* was described as having both creative as well as destructive powers which were dependent upon the vision of *Zeus* or the controlling intelligence. Further, as Plato suggested, this model is a model for a creative as well as a destructive human.

Entropy disrupted science even further because science now had to recognize that life responds differently to entropy than does the physical universe. Life has a negative flow of entropy in violation of the *Second Law*. The entropy which causes disintegration and separation of building blocks is actually being decreased by the process of living and evolving. Humans are in the process of becoming more and more in disequilibrium with their physical surroundings and can only survive by consuming other life forms.[84] All humans as well as all other life

---

[83] See Prigogine (1996).

[84] See Schrödinger (1992).

forms also have their own inner Law, energy and chaos similar to the original creation models.

Life is now starting to be described as a vitalizing force which couples and unites intelligences and reactions together and is even starting to be recognized as able to unite mind and matter. Not only does life coordinate all inner processes within the human body, it also integrates separate life forms together. An ant colony, for example, needs to be considered as a whole organism and has long been known to possess a collective intelligence far greater than any single ant including the queen (no brighter than any other ant.)[85]

Early Greek soldiers were reported to have rituals that increased their collective intelligence or unity (*henosis*) so that they could present a unified front to their enemies for better protection as well as power. People who have found a union in facing a common danger generally comment on how the entire group functioned as a unit without any outer organization or control and that the group was far stronger than the sum of its individuals.

In order to obtain some support for the coupling of mind and matter as well as between intelligences, it is useful to describe a bit more about the work done at the Princeton Engineering Anomalies Research Program, (PEAR.)[86] Their first success came when they were able to prove that the intention of an observer was able to produce a desired change in a precision machine in which small changes could be measured over time. The magnitude of change was very low such that if someone were to con-

---

[85] Vertosick (2002)

[86] See Jahn and Dunne (1989), Ch. 5.

centrate for a full twenty-four hours on a clock to speed it up, the clock would advance only about one minute.

I would explain the importance of their work as changing the balance of normal variations called "noise" in scientific circles. There is a random nature in how things operate together which, if pushed just a bit in one direction or the other, can greatly influence how things end up. For instance, I was able to convert the flow of heat into electrical power by slightly increasing the transfer rate of ions on one electrode compared to an opposed identical electrode in an electrochemical cell.

Chemists are well aware of how a very slight change in acid content can vary the resulting chemical reactions. Modern cars have more than doubled their engine life because of a slight improvement of lubricants. Princeton gave positive proof that "noise" can be altered and as a result can affect the brain as well as mechanical and electronic devices. It is also obvious that the PEAR group's results could not interest the news media or even those interested in the occult because of the extremely small immediate results.

The second efforts of the PEAR group consisted in demonstrating that a subject's mental impression of an observed scene could be mentally shared with a remote partner waiting and open for visual impressions of what was being observed. This project, like their machine control, was not attempting to demonstrate impressive psychic powers of "gifted" people, but rather to demonstrate that almost everyone is able to influence the physical world to a limited extent with his or her consciousness. A simple support for this concept is the belief that athletic teams have a "home team advantage" because of the physical presence of their fans. The results again are not dramatic; just as the PEAR subjects could not stop a clock, enthusiastic fans cannot assure victory to a home team. A person seeing a remote building or object could not transmit a "photocopy" image to a recipient, but could send an impression that would allow the later viewed scene to be identified.

This power reminded me of how people will only recognize something when they see it, such as during a police line up, or only being able to remember a name when it is seen or heard. Another great finding from the PEAR work was that impressions related to some interest were not limited by either time or distance.[87] In research I have often been amazed at my interest in some article which, even though of no immediate application, was important in my later work. Many lovers report how they had a mutual identification at first contact.

Many people when required to either give or receive needed advice are later amazed at its accuracy, and of course dreams and personal visions can open the door to evolution. Edwin Diamond gives an excellent summary of the power of dreams and describes an experiment demonstrating how dreams can search chaos for an answer.[88] Nevertheless, in all cases what is received can only be described as impressions such as those attributed to the *Muses* by the early Greeks. The brain must be used to verify and formulate the impressions.

This small inner power reported by PEAR can be directly related to the power of Love which, as in Plato's time, receives little praise even though most people have an

---

[87] Ibid., Ch. 1.
[88] See Diamond (2006).

impression of its potential power. Love is described as a continuous glowing ember residing within the heart which constantly controls each small step of life ultimately leading to the fulfilling of the heart's desire[89] (for evolution or self-destruction.)[90] *Eros* or Love is able to instantly release tremendous mental, physical, and metaphysical powers but only by a process similar to turning on a hidden switch if the oncoming step is sufficiently threatening. At that time a little nudge of consciousness can then stimulate a chosen inner organ to release a hormone which is then able to turn on the residual powers of the body to fulfill what has been placed in the heart.

With the above descriptions it is now possible to describe the early Indian theory of creation. The Indian philosophies are quite similar to Hesiod's model of creation, but with an increase in very useful details of the four elements of *Zeus*, Chaos, *Eros* and Time. *Zeus* becomes *Brahmā* for the personal inner source of Law; the heavenly *Zeus* becomes *Brahma*. The Greek Chaos becomes *Māyā* and is much more than a cosmic material dumping ground. *Māyā* has an added meaning closely related to the Biblical usage of the Greek *tamias* or *tameion* which means the dispensary of *Zeus* or the hidden treasury.[91] *Māyā* can be considered as storing the feelings of memories, the causes of memory as well as their effects in conjunction with the brain. *Māyā* is treated as a reference library, and for creative people it can be viewed as the source for the answers to problems being slept on. Others view *Māyā* as an exchange center between their inner soul and *Brahmā*.

---

[89] See Proverbs, 16:9 and Peck et al. (2006), Ch. 13.
[90] Plato *Phaedrus*
[91] Matthew 6:6, 24:26

In connection with *Māyā* there was also the process of sorting through everything which had been placed in *Māyā* called *saṃskāra* (which means to put together, or to form the mind.) For instance, *saṃskāra* is credited with a power of *Brahmā* which collects the mental and metaphysical elements for an individual at birth including the seeds for what is initially within the heart which makes every newborn unique. After birth, the power of *saṃskāra* is allegorized as the power of an inner god, who was equivalent to *Eros.* Loewi's dreams are an excellent example of how *saṃskāra* can be conceived as pulling from *Māyā* the elements of the answer as to how hormones communicate. This process was not conscious and is easily allegorized as being done by an inner god, the equivalent to *Eros*.

There is a further and surprising similarity in the Indian model of creation to that of modern science and that is with the "singularity" of creation. Science describes the concentration of the stuff of creation and energy into a point to produce the explosion as a "singularity". Sanskrit has a term for a similar type of a creative singularity which is *vajra*. A common definition of *vajra* is that it is the creative or destructive thunderbolt of *Indra.*[92] *Indra* can readily be compared with *Eros* and is described as controlling the various vital forces and powers within the body.

The early writings require the thunderbolts of *Indra* to cause a change and hence describe how *vajra* can be stimulated and controlled. In terms of individual creativity, an equivalent of the Big Bang must take place within the individual or *Indra* must release a thunderbolt. Science

---

[92] *Indra* later became *Śiva* known to most students of *hatha yoga*.

does not explain how the singularity could be initiated or directed in the cosmos; however, it can explain the near instant response of the body to threat or trauma.

The modern description of the secretion of adrenaline and the immediate stimulation of the body would certainly be described by the early Indian scientists as a *vajra*. Similarly, the vivid dream state or REM sleep is now known to be almost violent with the increased depth of breathing, the stimulation of the perineum, and the response of the brain which causes the "more real than real" dreams as well as the common precognitive-type dreams. Sudden changes in the body do not seem to come slowly, but with force or like a thunderbolt.

There is yet another comparison of the change of the unformed random chaos into the orderly cosmos of the universe to that within an individual. The earliest inner ruling gods were allegorized as containing the power of storms, wind, and fire. The inner stabilizing and growth processes were related to the inner rains of *Indra* which compares well with recent work with the curative powers of intense crying in which hormones are secreted in tears as well as the stimulation of the lower gut churning. Further, the work of Prigogine suggests that increasing the energy in chaos can result in creation and organization of the elements of chaos.[93]

This emphasis on increasing the energy within chaos is fundamental to the ancient concept of evolution which is in agreement with that of the modern astrophysicist. The use of increasing inner chaos to evolve was, no doubt, a practice in ancient Greece, but all that remains are statements

[93] See Prigogine (1996).

such as that of Plato stating that evolution required madness or chaos represented by the gods, *Apollo*, *Dionysius*, and the *Muses* well known through celebrations.[94] *Dionysius*, rather than allegorizing drunkenness as his detractors claim, was the power of initiating the release from the conditioned self. It should also be recognized that chaos was not the result of Gymnastics or exercise. Gymnastics fed the body, whereas madness fed the soul.

In general, many of the ancient practices for increasing vitality and evolution in life were based on increasing the interaction with life and all of its choices. Similarly, in solving problems, letting the mind scan chaos is far better than clinging tightly to a desired answer. Many people report that when they sleep on a problem, they can experience many scenarios before the answer seems to drop into place.[95] Touching chaos before finding an answer can also be expressed as how darkness preceeds dawn or how each cloud has a silver lining.

---

[94] *Phaedrus; Laws*

[95] See Wagner, et al. (2004).

Chapter Eight
Becoming Gods

Before discussing the process and methods of becoming a god, it seems necessary to describe what the ancients considered a god to be. The book *Cratylus* by Plato is primarily concerned with a discussion on this very subject. His discussion concludes that a cause of something is given a name which describes the effects of that cause. Plato considers the meanings of the names of a number of gods and demonstrates that the origins of those names describe particular phenomena. The ancients considered a god to be only a label or description of the source of some unexplained manifesttation.

It seems that the naming of metaphysical causes as gods is still in wide usage, as for instance in medicine when doctors explain that their patients suffer from *myalgia* (from *myo*: muscle and *algia*: pain). *Myalgia* is a symptom of many diseases or disorders. That is, *myalgia* is the effect of any of a large number of causes, yet the use of the word *myalgia* implies that it is the cause of the muscle ache itself. We treat *myalgia* as a god (or a demon) since it has some unknown metaphysical source which is able to overpower an individual. For another example, consider the names given to the source of menstruation in women such as the Intruder, the Friend, and the Curse. Or consider how our society fears the god of Depression which can control an individual's entire view of life. We give a name or label to those powers which we cannot otherwise explain.

Individuals become a source of power or a god when they become the metaphysical cause of some physical effect that would not have otherwise occurred. The ancients called such individuals heroes (from exhalation, “h” + the god *Eros*). Heroes are those who dash into burning buildings and snatch children from the very jaws of death or those who lead a political rally to defeat some powerful oppression. During the Dark Ages the populace was told that there were no inner powers or gods and that all causes came from heaven. There could be no such thing as an individual, object, or substance with god-like powers, but fortunately the world is gradually recovering from such ignorance.

In contrast, prior to the Dark Ages there was a great interest in the ancient world of how to become a god. For example, Aristotle quotes an earlier philosopher who said that an individual becomes a god by having an excess of virtue.[96]

In the modern world, Abraham Maslow asserts to the truth of this statement when he confirms that exceptional or enlightened people are strongly dedicated to some goal and whatever they do must be done well. Their chief characteristics are their creativity and freedom to see the world as it is or as it will become. He noted how they spontaneously do right, which is certainly a trait one would expect from gods. Maslow was also emphatic that the self-actualized or fully human person suffered from pain, sorrow and troubles.[97]

It was perhaps Maslow’s mention of sorrow and troubles which turned my attention to the Biblical story of the Garden of Eden which, with a bit of thought, seemed to be a very concise and universal statement of how to become a god. However, in order to do this, I found that I

---

[96] Aristotle *Nichomachean Ethics*
[97] Goble (1970), Ch. 3

had to use the original meanings of some of the Hebrew words rather than the popular interpretations. The story starts in the perfected Garden with the interaction between the resident Gods[98] and Adam and Eve. One resident God instructs Adam and Eve that they cannot touch or assimilate knowledge from its central source[99] or they will die on the spot. Later, Eve's foresight[100] tells her that if she acquires knowledge she won't die, rather her eyes will open, and she too can become as the Gods. Adam and Eve do acquire knowledge and their eyes do open and they do become as the Gods (at least to the extent of knowing good and evil.)

The story then ends with a resident God affirming Eve's insight, adding that there is a price that must be paid. Since Adam and Eve now are able to see the world as it is and with understanding, they will find that every creative act as a (mortal) god must be paid for with suffering and toil. Further, since they must leave the Garden, there will be enmity and a barrier between their inner source of knowledge or foresight and their reason or conscious thoughts.[101] They are also reminded that they are not yet full Gods, since they are as yet mortals. The resident Gods gives them clothes and bids them on their way. Adam and Eve then act similar to the Gods of the Garden as they too become rulers and caretakers of their livestock, crops, children, etc. After leaving the Garden they have access to *Yehovah* (the Existent) and are no longer subject to *Yehovah Elohiym* (Existent Rulers), and even though they

---

[98] *Yehova* from *hayah*: "existing" + *Elohiym*: "deities or rulers"

[99] *atsah*: "hard, material source or tree"

[100] *nachash*: prognostication: pro: "supporting" + gnosis: "intuition." Also translated or allegorized as a snake or serpent; see the snake of *Asclepius,* Ch. 6.

[101] See Pascal (1993).

are mortal, they are able to cause changes leading to even greater creations than the creation of the Garden.

The story of the Garden of Eden exemplifies two major traits found in individuals. The first is that of being a caretaker or god with special powers and the second is that of being taken care of. The majority of people have experienced the role of being a god at least momentarily upon facing a demanding need with a strong dedication such as experienced by parents, teachers, physicians, policemen etc. Maslow calls these events "peak experiences" and the ancients describe them as being created by the nourished or activated inner powers or gods. However, those who manage to fully live as mortal gods are but a small percentage of the population. Maslow numbered them as far less than one percent. The recently discovered *Gospel of Thomas* estimates their number as one in a thousand or two in ten thousand. The reason for the low numbers is, no doubt, that the majority of people lack sufficient vitality and self-identity to accept being a god or accept having special powers to direct their own lives much less someone else's.

The next question is what makes the difference between the ruling gods and those desiring to be ruled or taken care of? In *Toward a Psychology of Being*, Maslow argues that research is needed to study the god-like or self-actualized with at least the same effort spent studying those dependent upon the gods or institutions. He also made an interesting statement that such researchers must have their eyes opened to "all sorts of basic insights, old to the philosophers but new to us."[102] However, even with the rapid rise in the knowledge of hormones, which surely

[102] Goble (1970), p. 24

provides a starting base for research, such a study would, no doubt, be extremely costly and it is very unlikely that any pharmaceutical company would support research which would diminish the need for its medications.

Maslow's statement that the answer is old to philosophy suggests that he thought the answers might already exist in the ancient philosophical writings as basic insights rather than hard science. What is very supportive of his assumption, but also furnishing a solid scientific basis, is that the ancient philosophical writings do provide answers as to how to gain powers. It certainly seems logical that the ancients were quite capable of controlling their inner powers since they created civilizations unequaled today and without the support of modern technology such as modern transportation, telecommunications, power tools, medicine, computers etc.

For a start in searching the old philosophies, consider the medieval Indian document summarizing a much earlier universal science of developing and controlling inner powers. The document is the *Haṭhapradīpikā* (also called *Haṭha Yoga Pradīpikā*) which can be literally translated as *Violent Bestowing of Union*[103] but which is widely interpreted in Western yoga groups as teaching "gentle stretches."

The first chapter[104] in Book 1 describes a practice beginning with the pounding of the buttocks against the floor which release the rains of *Indra* (drainage of the eyes, nose, and mouth.) This is followed with sitting on the foot in the Great Posture, *mahamudra*, to bring changing pressure against the middle of the perineum while leaning

[103] *hatha*: "violent," *pradi*: "bestow," *api*: "annexing, uniting"
[104] Verse 1:35 from Book 1 of the *Haṭhapradīpikā*

tightly forward. In this position, the guts are churned using a tense and forced exhalation. The verse assures the reader that this practice will open the door to enlightenment. This is, no doubt, a Gymnastic that Plato would call food for the body.

The scientific truth of that *Haṭhapradīpikā* verse was proven in my mind when I could recollect performing a very similar "violent bestowing" as a very young child. I can remember one session very clearly which started when I had been falsely accused of breaking some rule and felt unbearable frustration when no one would even listen to my defense. I remember being alone and sinking deep into depression as my world shrank to include only the accusation. However for some reason, I yielded myself to some inner force as I sank even deeper into the feelings.

The feelings overpowered me and I dropped to the floor in a corner of the room. Something mysterious then seemed to take control of my body as I felt myself doubling up so that my thighs were pressing tightly against my belly. I was pressing so hard that I could only gasp as I tried to press the inner deep seated pain down through my body to the floor using very forced exhalations. Then I felt my belly churning strongly with what could have been interpreted as attempting to purify the interior of my body. Fluids were pouring out of my eyes, nose, and mouth such that I could only gasp for air as they also seemed to be washing something away. I remember the rise of a strange constriction in my chest which felt related to what was happening in my belly.

However, as I tried to respond to the sensation by pounding, squeezing, grasping and finally pressing and rubbing, the feeling in my chest became more and more pleasurable. The pleasure seemed to flow down into my groin without sexual arousal. My perineum became extremely sensitive and actually demanded that I respond with a strange form of rocking which resulted in generating a rising sense of pleasure which was only increased with the downward flow of pleasure and the motion in the buttocks. The rising pleasurable vibratory sensations had somehow completely replaced the overpowering pain and frustration that started the whole collapse. I then felt like I had awakened from some dream which continued to hold me in its beauty and joy. I can remember wondering about my original frustration and pain, but I could not reproduce it or even give it any meaning in my new world.

I am quite sure that many people would identify the above episode as "sobbing" or perhaps as "crying your heart out." As I got older I learned about another side of sobbing, probably when my parents caught me slobbering, gasping and squirming and indoctrinated me with the warning that sobbing was violent, dangerous, harmful, as well as sinful to do. Hence, I had to abandon my "sinful" practices as almost all other growing children must do and learn to get control of myself. However, I compromised and learned to internalize the wonderful ending feelings.

As an adult I developed the method described earlier of rocking slightly on a bulging cushion and pressing down with a tight control exhalation. I found much later that even this simplified motion is mentioned in the Indian *RigVeda* [105] when it states that everything can be gained if one sits pressing downward and moving the hips forward and back like a woman.

---

[105] 1:28:3

Later in my studies I became convinced that the sobbing I had experienced was a natural and universal body and mind response, and it would, no doubt, be a starting point for early research such as suggested by Maslow. What was surprising was that even though Western writings referred to practices that produced the god nature, I could not find any descriptions of what the practices actually were.

That the practices had an element of violence is perhaps evidenced by Proverb's insistence that paddling a child is the best method for correcting a child's misbehavior. There has been some research into the value of crying[106] and laughing[107] with the observation that intense belly activity results in the emission of hormones in the tears.

Maslow was certainly correct in assuming that the ancient basic insights are difficult for the modern world to see. Since I could gain little from existing Western writings about the "secret" practices used to become a god, I found that I had to use the writings of India. I also determined that the ancient terms were far more detailed and accurate than current terms.

For instance, the ancients obviously used the term *soma* in referring to hormones, but their writings relate far more about the generation and control of *soma* than can be found in modern endocrinology. The term *yoni* describes characteristics of the perineum that are largely undeveloped in the modern world, and the word *hridaya*, normally translated as "heart," has nothing to do with the beating heart in the chest. Because there were so many other unique terms, I decided that I would present the literally translated docu-

---

106 See Solter (1984) and Frey (1985).

107 Provine (2000)

ments and let the reader find their own interpretations. The literal translations seem to provide a clear description of the methods, but certainly the effects are highly personal.

Chapter Nine
The Physiology of the Breasts and Perineum

In order to understand the teachings about the ancient method of controlling your inner hormones, it is necessary to understand the physiological interactions of your breasts and perineum. This is not a simple task since we are all taught from childhood that our center of feeling and action is in the beating heart. There is almost no scientific support for such a statement and perhaps it persists because science has not yet opposed the popular "poetic" view.

It should be noted that science has no tools to directly measure "feelings" since they are by nature metaphysical. Because of this, science is actually reverting back to the ancient anthropomorphic type descriptions which got the old science labeled as pantheistic superstition.

It is significant that both breasts[108] and the perineum[109] are now described as tending to atrophy because of non-usage. The muscles of the perineum have been of such a low concern that the names of the muscles are not agreed upon and the logic for their original names lost. One large barrier to the ancient wisdom is the sexist view of breasts. Both men and women have breasts and nipples, the difference in sizes are primarily due to hormonal or physical stimulation of the basic structures.

Rather than attempt to counter present popular beliefs and concepts, it seems much easier to start with recent scientific discoveries and use them to support the ancient sci-

---

[108] See *Gray's Anatomy*, p. 1038 *The Mammary Glands*.
[109] Resnick, et al. (2003)

ence even though this route may seem quite indirect. I will at least try to make it interesting.

Let me introduce the modern work with a well-researched paper with the horrific title: *Modulation of Neuroimmune Parameters During The Eustress*[110] *of Humor-Associated Mirthful Laughter* by Lee Berk and associates. This can be interpreted as stating that a "good" stress obtained with laughter can alter the basic hormones of the body which will be seen to be in direct support for ancient science and this book. The authors' equate their findings with Proverbs 17:22, "A merry heart doeth good like medicine, but a broken spirit drieth the bone." Berk and associates end by stating that, "...perhaps medical science is beginning to catch up to intuition."

One very important point of their paper is how the body can break free of the conscious brain's control of the release of hormones. During stress (both good and bad) the body's sensors and brain activate an inner control called the hypothalamus-pituitary-adrenal axis in which hormones generated in the head activate the release of steroid hormones from the lower abdominal adrenal glands.

However, once started there is the possibility that the hormones produced from the adrenal glands can cause even further production of the steroids without the necessity of the upper control hormones. This ability of some receptor sites of hormones to start the generation and release of the same hormones once they are turned on is an important function of the body. This is one method to switch the control of the body's reaction from the central nervous system to the autonomic control system.[111] If unchecked,

---

[110] Greek, *eus*: "good or positive" (eustress: "productive stress")
[111] See Ch. 3.

this shift in control results in the continuing unhealthy hypertension or stress syndrome encountered by many modern individuals as the adrenals continue to secrete steroid hormones.

Berk et al. carefully measured the hormones being produced in the body before, during, and after laughter and described how laughter could return the body and mind to an original balanced state called homeostasis. They support the ancient method of using the mind to control hormones with their observation that anticipation of laughter could produce the same results as laughter.

Earlier research into crying by Frey[112] demonstrated that the power of crying could produce eustress or could also counter stress and produce homeostasis. This balanced hormone state was obtained in large part by the ability of the body to remove excess stress hormones through tears. Unfortunately, Frey's work gained even less attention and support than did Berk's, no doubt, because crying is less acceptable in our modern society than is intensive laughter. There likewise is no profit potential associated with instructing individuals to cry or laugh.

The ability of the body to purify itself with tears can explain the Indian inner allegorical god *Indra* who used his clouds to produce purifying rain. *Indra* in many ways is the early description of the powers of the hormone histamine. Histamine is well known today as the cause of runny noses, and watery eyes which are equated with fighting pathogens or airborne irritants. It does this by increasing the permeability of blood vessels which leads to the swelling of tissues.

---

[112] Frey (1985)

There has been a large amount of related work concerning obtaining homeostasis with the prolactin hormone, which is now considered to be a multifunctional hormone rather than just a hormone which stimulates milk production. Prolactin, like laughter and crying, promotes the return of the body to homeostasis by its ability to change other hormones of the body to produce a balance. Prolactin is a very interesting hormone and is known to be dependent upon a number of external forces such as light, sound, odors, sexual stimulation and other hormones such as oxytocin. To this list I will add physical stimulation of the breasts which produces oxytocin.

At the time of the postpartum for a mother and child, nipple stimulation is well known to result in the rapid allostasis[113] of the body or its return to homeostasis. The stimulation initiates the repositioning of all of the muscles and nerves used during pregnancy and delivery. That nipple stimulation releases hormones is evidenced by the ability of muscles to be moved which cannot normally be intentionally moved. This connection between the lower muscles and the nipples is also noted when nipple stimulation can speed up the lower muscle activity to speed up delivery of the child.

The link between the deliberate stimulation of the nipples and the activation of normally quiescent perineal and lower abdominal muscles during child delivery strongly suggests that the nipples are the cause for the tightening and contraction of the perineum on stepping into a cold shower or the sudden turmoil in the guts and the holding of the breath when facing danger. This suggestion becomes more than a suggestion when it is remembered

---

[113] Greek: *allo*, "alternate" + *stasis*, "state"

that the nipples are active outside of nursing. The nipples are known to harden and swell when they get cold, and, furthermore, we are often aware that our nipples seem to swell and harden in response to our thoughts. This is, no doubt, a main reason why women choose to keep their nipples covered and their lips painted — to hide a similar emotional swelling. (Men, of course, have a similar problem, but manage to hide their swellings with clothing and with grimacing or clenching the jaw.)

Strong emotions are generally first physically experienced with three different strong reactions of the breath. It can be as if you are hit in the chest which forces out a strong exhalation, a strong inhalation as you suddenly have to suck in air, or as if the chest locks up and you find great effort to breathe. In order to explain these changes in breathing and the following quite varied emotional reactions, it is necessary to review the perineum.

The chief muscles of the perineum are shown in the Illustration on page eighty-seven. Of special interest is the central connection of the longitudinal *bulbospongiosus*[114] (B) and transverse muscles (D) which also connect to the underlying *pubococcygeus* muscle (E), a lower subdivision of the *levator ani* muscle which supports the whole abdominal cavity. The *bulbospongiosus* muscle can be constricted to stop the flow of urine and is felt while tightening the anal sphincter. This tensioning is normally done without disturbing the *levator ani* or the lower abdomen.

---

[114] The *bulbospongiosus* is also called: *accelerator urinae*, *ejaculator seminis*, and the *bulbocavernosus*. In the past it was also called the *pubococcygeus* muscle.

The *bulbospongiosus* muscle is, however, capable of stimulating the autonomic system which can then override the conscious control by the brain which is, no doubt, why the perineum was labeled as the central control for the body (*peri*: central, *neuma*: control) necessary to protect the body. One of the first controls associated with the stimulated *bulbospongiosus* muscle is the ability to activate the *levator ani* which is normally incapable of being consciously moved.

Both the *levator ani* and the *bulbospongiosus* muscle respond to the nipples and breasts as evidenced by the strong activity of the *levator ani* during the rise of intense emotions when deep gut churning and even vomiting can occur. Further, the *bulbospongiosus* muscle is able to swell in various sections of its length to change its stimulation of the autonomic system.

The Muscles of the Perineum

Greek: *peri* "center" + *neuma* "control"

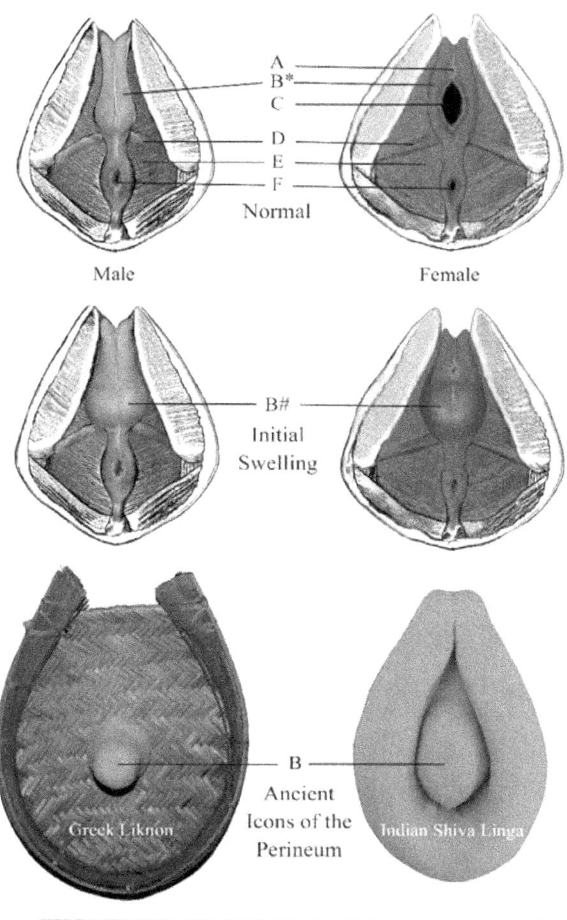

A: Urethra B*:Bulbospongeousus muscle (Also called accelerator urinae, ejaculator seminis, bulbocavernosus, and often pubococcygeus muscle.)

B#: Bulbospongeousus muscle (Also called Tongue of Agni, kanda, phallus.)

C: Vagina D: Transverse muscle E: Pubococcygeus muscle F: Anus

Chapter Ten
The Ancient Practices

The ancient self-developmental practices can be introduced by the following description based upon Chapter Nine and the paper by Berk (2001). The ancient self-development practices for obtaining homeostasis were based upon first developing the body to obtain a powerful eustress and then shifting the former undesired homeostasis with an increased energy and control to effectuate allostasis.[115]

This can be simply stated that the ancient writings teach that the first step is to overcome the socially induced tensions and stress by means of the generation of different hormones. This is done with special practices that teach how to turn off the inherently low-energy, judgmental brain chatter by adding to the body and mind an added high energy good and productive stress. The activated brain is then taught how to control the awakening inner *Me* as it restates or enforces the basic desired goal of life. The individual then learns how to transform the desired goal into what can only be described as being as real as a memory.[116] The final step is to bring the memory into the present and then physically manifest it which forces the release of supportive hormones to reproduce the imagined role and world.

The ability of the mind or behavior to change inner hormones must be recognized as a fundamental concept in both the ancient as well as modern science.[117] This con-

---

[115] An atypical change in hormone flow, see Ch. 9.
[116] See Ch. 2, *Mnemosyne*, memory or mother of the Muses.
[117] Adkins-Regan (2005), Ch. 1.

cept is, however, little known by the average individual who has learned to be dependent upon drugs to change their inner body. It is also important to understand that hormones have the ability to maintain or even increase their own strengths once they are released and bonded to receptor sites. These powers of hormones are included in the ancient description of hormones as *soma* and the powers as gods.

If problems are encountered in understanding the following texts, it might be helpful to think of the versatility and changeability of young children, which were after all, surely the source of the ancient science. Thinking of children is certainly helpful in thinking of the power of breasts to feel and to produce the nature of feelings as thoughts. Children use this ability of the breasts to become fearful or angry by first describing something scary or threatening and making it real to the mind. When it becomes real to the mind then the breasts respond to it creating the desired emotion along with the rise in vitality of the resulting hormone release. This process will assist in understanding the powerful ability of *mantra* and *mudra* to change your world described in the *Rudrayāmala*.

The Sanskrit writings are very much centered upon the swelling of the *bulbospongiosis* muscle of the perineum. The swollen bulb is called the *jalgula* or covered bulb in the *RigVeda* and later the *kanda*[118] or *lingam* (meaning a mark or sign which proves the existence of something.) The winnowing basket is well explained in many sources as depicting the inner motion of the abdomen as it purifies the forces of the body similar to the separating of grain from chaff.

---

[118] Defined as a fleshy protuberance from the perineum.

Chapter Eleven
The *Haṭhapradīpikā*

Introduction

The *Haṭhapradīpikā*, also known as the *Haṭha-yoga-pradīpikā*, is a very powerful ancient book which can be described as teaching what the Western world would call how to "gird your loins" or how to put on or gain a more powerful body and mind. Written in Sanskirt, the *Haṭhapradīpikā* uses the word *bandha* instead of girding, but *bandha* has the identical meaning of "binding together" or "putting something on."

About five hundred years ago an Indian scholar calling himself *Svātmārāma*[119] assembled much earlier Sanskrit verses of undetermined age into a document he named the *Haṭhapradīpikā*. His document supplied the original verses, his rendition, and his commentaries.

That the *Haṭhapradīpikā* even exists is exciting, since the Western world has writings recommending girding, bracing, or steeling yourself, but has no surviving writings describing how it is done. Nevertheless, despite the lack of information about girding, people seem to instinctively know something about the power of girding and can identify with it. For example, many people in facing life's problems often describe the need for a strong resolve or dedication and then report the uncontrolled tightening and activating of something deep in the guts or the perineum to make it happen, which according to the *Haṭhapradīpikā*, is the beginning of girding the loins.

---

[119] *sva*: "one's own," *ātma*: "soul," *ārāma*: "a place of pleasure"

The *Haṭhapradīpikā* and other Sanskrit writings survived the book burnings of the West by what appears to be the deliberate addition of false translations and commentaries to the original documents to satisfy the ruling British colonists.

The original technical Sanskrit used in describing the science of the body and mind is similar to modern concentrated scientific writings which almost everyone can read but only a few knowledgeable individuals can understand. Most of the technical Sanskrit writings include a requirement that their content must be kept secret. This is easily done in a technical manuscript by adding a false meaning of the critical terms which can misdirect the uninformed or uninspired reader. My guess is that the rendered translation of verses 3:33-34, that a seeker of truth needs to slice the base of his tongue so that the tongue can reach up into the sinuses, was such a misdirection. Censors would surely not condemn such an idiotic statement, and in fact if they were of another culture, they would openly endorse such statements to prove the superiority of their culture (which was quite often done.) However, if the censors had any scientific knowledge and the diligence to study a technical document, the original teachings could have been readily obtained by using this knowledge and accurately translating the Sanskrit.

Because of the socially conforming commentaries and translations, any existing concepts of the book or *Haṭha Yoga* should be set aside. Instead of implying gentle exercises and stretches, the title literally means: *Haṭha*: “force,” + *pradīpikā*: “a treatise on.” The literal translation is in full support with this title.

The following text is given in the following format:

1. the verse number and a twenty-first century rendition,
2. the literal translation,
3. the common popular interpretation,
4. the transliterated verse,
5. a list of the original Sanskrit words and roots, and
6. a comment to assist in further definition of the terms and concepts.

It should be noted that the Sanskrit of the *Haṭhapradīpikā* requires a basic understanding to "decode" the meaning since there are very few verb declensions or noun cases common in other types of Sanskrit documents. Further, some words are joined together with an alteration of both the beginning and ending letters[120] that allows the hiding of a meaning under a constructed word built upon words with different meanings.

Fortunately, there is a very clever built-in guarantee that a translation is correct when complete correspondence can be found between the translation and the literal words, since there are no unessential or undefined words. This is also true for modern technical scientific writings. Historically, a dedicated individual would find an experienced "guru" to assist in translating the text and understanding the content, and thus the complexity of the Sanskrit would be of no concern. The technical nature of these verses has also served to carry the ancient knowledge to the modern world, much as the early scientific writings of the West are understandable centuries later for those who are dedicated to searching.

---

[120] Rules of *Sandhi*

It might be useful to review the critical organs of the body described in earlier chapters. The location of the ancient heart is, no doubt, the sacrum which contains a large amount of gray cells and connective "horse hairs."[121] This bone is located, as claimed, in the center of the individual if the body is measured from the tips of the toes to the extended fingers. The sacrum is known to have large neural connections to the center of the perineum or *yoni*.[122] The swollen bulb is obviously referring to the *bulbospongiosus* muscle (in both sexes). This name describes the swelling and sponge-like or moldable nature of the muscle.

Although modern physiology is aware of these organs, there is remarkably little reference to them, even though as mentioned above, the atrophied perineal muscles found in aging bodies are now known to contribute to low vitality and incontinence.

The ancient world, however, was well aware of the *bulbospongiosus* muscle. The best example is obtained from the artifacts of the very advanced civilization of the Indus Valley. Even though they left no writings, they did leave what is now called the *Shiva*[123] *Linga* icon (See Illustration pg.87) which depicts the swollen middle section of the *bulbospongiosus* muscle protruding from a pudendum[124] symbolizing creative power.

---

[121] Gray's Anatomy

[122] *yoni*: Sanskrit: "place of origin, source of generation"

[123] *Shiva* is a later name for *Indra*, a name used for the inner creative power or soul.

[124] The protrusion is also falsely explained as a "prolapsed uterus."

The swollen protrusion[125] of the *bulbospongiosus* muscle is described often in the *Haṭhapradīpikā* and other Indian writings, and its stimulation from changing pressure, squeezing or inner movement is a beginning *bandha.*[126] It can also be perceived to be the "phallus" in the Greek *liknon* of the early Greek Dionysians (verse 3:8). The sponge-like nature of the *bulbospongiosus* muscle explains why it is sometimes depicted as a pillar-like protrusion and other times like a swollen bulb depending upon the nature of the pressure placed upon it and its excitation. It is, no doubt, the ability of the *bulbospongiosus* muscle to swell and move in different sections that it was compared to a tongue in a lower mouth.

It is helpful to read the ancient text as the early descriptions of the feelings and reactions of controlled hormone production. The ancient Indians called the flow of hormones *soma* or *amrita*. It is interesting that the rains of *Indra* or tears have only recently been explainable as releasing and balancing the inner hormones of the body.[127] Similarly, the inner feelings of a rising tongue of fire or energy known as the tongue of Agni is the release of serotonin and the adrenal hormones which is described as a rising, intoxicating flow or similar to a "drug rush."

The following excerpts from the *Haṭhapradīpikā* do not include any verses from the final section which describes the *nādam* or the inner sounds or vibrations of the inner energy. Although most children sense the *nādam* in their activities, it is suppressed with aging and considered to be an illness called tinnitus or "ringing in the ears" by

---

[125] Called variously: *jal gula*, *kanda*, *khaṇḍa*, *svayambhu*, *yonyarśas*. Also listed as the garden of *Indra*, *kanda sara*.

[126] See *RigVeda* 1:28, in Ch. 12.

[127] See Ch. 9.

modern medicine. Although no one knows the source, nature or function of tinnitus, it is being "treated" with sedatives, which suppress its source or vitality.

The *Haṭhapradīpikā* states that ringing in the ears is associated with beginning *bandha* and further mastery brings awareness to its source in the belly. This final section is descriptive only and will be experienced by those who are able to effectively gird their bodies and minds, and hence I deemed it not necessary for this book.

Text of the *Haṭhapradīpikā* – Book 1

1:27 Severe diseases are destroyed with proper sitting while pressing the swelling bulb of the perineum and then churning. Enlightenment and spiritual gifts are obtained with regular and constant practice.

Literal: Severe diseases are destroyed by means of churning with proper sitting, the stimulated cavity, and the small protuberance of the bulb. Regular and constant practice pressing the bulb awakens, illuminates, and brings forth spiritual gifts.

Popular: This *matsyendrasana* (which increases appetite by fanning the gastric fire), is a weapon which destroys all the terrible diseases of the body: with (daily) practice it arouses the *kuṇḍalinī* and makes the moon steady in men.

matsyendrapīṭhaṃ jaṭharapradīptiṃ
pracaṇḍarug maṇḍala khaṇḍa nāstram /
abhyāsataḥ kuṇḍalinī prabodhaṃ
candrasthiratvaṃ ca dadāti puṃsām // 1:27

matsyendrapīṭhaṃ: the	***abhyāsataḥ***: constant practice.

posture for churning by means of inner powers. ***matsya***: ***math***: to churn. ***sya***:by means of. ***indra***: inner god of fluids. ***pīṭhaṃ***: a particular posture of sitting.
jaṭhara: interior of anything, cavity. ***pradīptiṃ***: radiant, stimulated.
pracaṇḍa: terrible. ***rug***: illness.
maṇḍala: ***maṇ***: ***maṇi***: jewel, protuberance. ***dala***: a small shoot, expel.
khaṇḍa: bulb, protrusion.
nastra: destruction. /
kuṇḍalinī: ***kuṇḍa***: pot, bowl shaped container. ***kanda***: bulb. ***linī***: ***lina***: pressed together, resting on.
prabodhaṃ: opening, awakening.
candra: illumination. ***sthiratvaṃ***: firmness, constancy.
ca: and.
dadāti: gift.
puṃsām: spirit. // **1:27**

Comment: This verse follows verse 1:22 (not printed) in which the anus is pressed while sitting on the heel as a starting practice, since initially the perineum is largely unresponsive. This verse then adds churning and the pressure (from the heel) against the small protuberance of the perineum which becomes a bulb. This is then a restatement of verses 1:28:1-4 of the *RigVeda* (Chapter Twelve.)

Churning requires initially both an upward pressure from the heel and a downward pressure from the lower abdominal and sexual muscles which can be initially physically tightened and loosened so that a rising and falling motion like a churn is obtained. The following verses describe how the bulb evolves and the churning becomes automatic with the usage of the churning rods.

The Sanskrit uses the word "foot" or "heel" in a generic way, since any object capable of providing pleasurable pressure against the *yoni* can be used. There is an advantage, however, if the pressure can be changed during a practice to continually increase the pleasure.

The word *kuṇḍalinī* is used in two different ways in the book. The first is as a compound word describing the pressing of the *kanda* or bulb, and the other is as the resulting reactions from that pressing. Thus, the term *kuṇḍalinī* names the cause as well as the effect of churning.

This verse's reference to enlightenment can be related to an experience of intense insight or intuitive knowledge that follows a major threat or danger. This is generally associated with unusual feelings and motions in the lower abdomen and sexual region as well as the pressing or clenching of the center of the swollen perineum. Another example is the experience of special or creative insights produced during some vivid or REM dreams. During this type of dream there is considerable motion in the lower abdomen and stimulation and noticeable swelling of the perineum with labored breathing. Children use the natural restoring sitting position with intense sobbing with the legs pulled tightly against the body to bring pressure to the lower abdomen.

The ancient world generally described the individual as having two different sources of intelligence and control: the brain and the gut. The head and brain was associated with heated action or the physical doing and was related to the sun. The lower source of intelligence was hidden, coming up as feelings from the darkness, and hence was associated with the indirectly lit moon or glow.

The following verse gives the basic concept of how the inner hollow in the perineum is to be pressed and a little more about the inner power called *Indra*.

1:35 Sit with pressure on the *yoni* or the active center of the perineum, exhale strongly, shake, churn and pound against the source of pressure. This can be used to explore the source of carnal feelings which can open the door to obtain enlightenment by means of the released inner power.

Literal: Take the position of steadily pounding and churning the *yoni* vigorously with the foot placed in different places. Fixed and bound with the power of *Indra* using the liberating exhalation and the appearance of shaking, the conquering enjoyable carnal feeling is observed and maintained, which opens the door to breaking through and producing enlightenment.

Popular: The *siddhasana* (is now described). Press the perineum with the base of the (left) heel and the (other) foot firmly above the penis (or pubis). Keep the chin steadily on the breast. Remain motionless with the sense organs under control and with steady vision look at the spot between the eyebrows. This is called *siddhasana*; it throws open the door to emancipation.

yonisthānakamaṅghrimūlaghaṭitaṃ kṛtvā dṛḍham vinyaset meḍhre pādamathaikameva hṛdayekṛtvā hanuṃ susthiram / sthāṇuḥ saṃyamitendriyo' caladṛśā paśyedbhruvorantaraṃ hyetanmokṣa kapāṭabhedajanakaṃ siddhāsanaṃ procyate // 1:35

yoni: ***yoni***, ***sthāna***: position. ***kam***: to strive after. ***aṅghrim***: ***aṅgh***: set about. ***mūla***: firmly fixed root. ***ghaṭitaṃ***: attempting, produced.
kṛtvā: causing.
dṛḍham: firm.
vinyaset: to place in different places.
meḍhre: ***medhya***: full of sap or pith, vigorous.
pāda: foot, pillar, support. ***math***: to churn. ***aika***: related to one. ***mev***: to serve.
hṛdaye: seat of feelings. ***kṛtvā***: causing.
hanuṃ: ***han***: to strike, pound.
susthiram: firm, steady. /
sthāṇuḥ: fixed.
saṃyamita: bound. ***indrio***: power of *Indra*.
cala: moving, shaking. ***dṛśā***: to become visible.
paśyed: ***paṣya***: beholding. ***bhruv***: ***bhṛ***: to maintain. ***rantaraṃ***: ***ranta***: to enjoy carnally. ***ram***: enjoy carnally. ***tara***: surprassing, conquering.
hy: ***hi***: to assist. ***etan***: breathing out. ***mokṣa***: liberation.
kapāṭa: door. ***bheda***: opening breaking through. ***janaka***: producing.
siddha: perfected, enlightened. ***asana***: posture.
procyate: ***pracyu***: to move proceed. // **1:35**

Comment: This verse follows 1:22 and 1:27 and the development of the perineum for both sexes from a sensitive anus to a swelling perineum and now to the development of the *yoni* in the center of the perineum.

The *yoni* is within the center of the perineum for both sexes and requires considerable effort and time to develop its full potential, particularly since it is normally suppressed with constant tension. Stimulating the *yoni* is for the release of *soma*, or the generation of an energy or food called *soma*, to be ready to respond to the needs of the body. This generation is often described as vitalizing the body by getting the guts excited, churned or in an uproar.

The awakening of the carnal feeling associated with the *yoni* is, no doubt, found by women massaging a sensitive spot called the G Spot[128] with a finger. This spot is initially found as existing between the vagina and the pubic bone. This spot can be found by men by inserting their fingers into the same location by starting from the scrotum. The following verses describe how the *yoni* can be directly stimulated and the *bulbospongiosus* muscles developed to become swollen, soft, compliant and supportive of the inner swelling such that the swelling extends well beyond the perineum as the bulb or *lingam*. (See Illustration pg.87.) The process takes weeks if not months to produce significant swelling, but is always rewarded with ever increasing carnal feelings. The usage of the word carnal also includes how the *yonis* of lovers can be coupled in a mutual *bandha*.

1:56 Sitting and breathing in different ways allows the manifesting of some desired image or *mudra* of the self, and as a result of the inner union of forces, a sound rises from the belly called the *nadam*.

---

[128] Ladas et al. (1982)

Literal: Thus, sitting and restraining the breath in different ways causes the manifesting of a desired self image or *mudra*. Then, because of the force, an inner sound of uniting rises from the belly.

Popular: *Asanas*, varieties of *kumbhaka*, the positions called *mudra*, the concentration upon the inner sound comprise the sequence of *hatha yoga*.

āsanaṃ kumbhakaṃ citraṃ
mudrākhyaṃ karaṇaṃ tathā /
atha nādānusandhānamabhyāsānukramo
hathe // 1:56

āsanaṃ: ***asan***: sitting posture. ***kumbhakaṃ***: water pot, nonbreathing. ***citraṃ***: different ways. ***mudrā***: sign, image. ***khyaṃ***: make known. ***karaṇaṃ***: doing, causing. ***tathā***: thus. /	***atha***: then, now. ***nāda***: loud inner sound. ***ānu***: to sound. ***sandhāna***: uniting. ***mabhya***: ***nābhya***: sprung from navel. ***as***: to take place. ***ānu***: to sound. ***kramo***: to advance. ***hathe***: in ***hatha***, in the force, violence. // **1:56**

Comment: This verse reminds us that body language and the force of the breath project an image of the self. As an example, the depth of breathing and the lowering of the pitch of the voice are characteristic of someone assuming authority. The sound rising from the belly is an introduction to inner sounds which are increasingly heard as the awareness shifts from what one thinks to what one feels. The control of what one wants to be, or the *mudra*, is discussed in detail in Chapter Thirteen.

The rising sound, called tinnitus or ringing in the ears, has no physiological explanation and is generally first experienced during a demand on the body such as in an illness or intense muscular effort. A survey of children has yet to find any child who does not hear the sound. The sound is "treated" by physicians with a sedative which reduces the energy of the body. (As a note: I tried to study the *nadam* with electronic equipment and could not find any measureable sound nor any correlation between an outside pitch and the inner pitch.)

Text of the *Haṭhapradīpikā* – Book III

3:8 The secret of the winnowing basket must be found with perseverance which is like the removal of chaff during winnowing or like the motions of a female body during coitus.

Literal: Endeavor and persevere to obtain the valuable secret of the wicker basket. In such a manner a small winnowing basket separates chaff in the same manner as a female body in coitus.

Popular: This should be kept secret like a casket of precious gems. It should not be spoken to anybody, as in the case of intercourse with a well-born woman.

gopanīyaṃ prayatnena yathā ratnakaraṇḍakam /
kasyacinnaiva vaktavyaṃ kulastrīsurataṃ yathā // 3:8

gopanīyaṃ: secret. ***prayatnena***: persevering effort. ***yathā***: in such a manner as follows. ***ratna***: anything valuable. ***daraṇḍ***: a basket of bamboo wickerwork.	***kasya***: Greek *liknon*. ***ka***: ***kad***: to remove the chaff or husk of grain. ***sya***: winnowing basket. (or ***ka***: little. ***sya***: winnowing basket.) ***cinna***: ***china***: taken away or out of. ***iva***: in the same manner.

akam: to endeavor to obtain. /	***vaktavyaṃ***: responsible for. ***kula***: abode, body. ***strī***: female. ***surata***: coition, amorous. ***yathā***: in such a manner as follows. // **3:8**

Comment: The meaning of this verse is supported by the Greek *liknon* (See Illustration pg. 87), a small winnowing basket the same size as the lower body containing a cloth-covered mound (See Chapter Six.) The comparison to the female sexual motion is similar to the rocking motion described in the *RigVeda* verse 1:28:3 as stimulating the lower bulb. I also like the Chinese description that the inner motion is like the mating of a dragon and a tiger in the belly.

This verse describes the basic motion of girding the loins or the gut reactions during intense emotions as the guts churn and generate hormones or vitality. The relationship to the elimination of chaff during winnowing is obvious as the negative aspects of emotions are "tossed away" in the wind of deep exhalations, leaving only the cleansed soul and body energized to do what the heart decrees.

3:9 With pressure against the *yoni* and with effort, the center is squeezed for the most pleasure. The source of pressure is held firm against the swollen active protuberance.

Literal: By means of a root support and striving, the *yoni* is squeezed for pleasure. The support is placed against the controlling swelling and held firm.

Popular: Pressing the perineum with the left heel and stretching out the right leg and grabbing the toes.

pādamūlena vāmena yoniṃ saṃpīḍya dakṣiṇam /
prasāritaṃ padam kṛtvā
karabhyāṃ dhārateddṛḍham // 3:9

pādamūlena: ***pada***: the foot, base. ***mula***: root. ***vama***: striving after. ***yoniṃ***: center of procreative energy. ***saṃ***: together. ***pīḍya***: ***pīḍ***: to be squeezed or pressed out. ***dakṣiṇam***: that which is right, pleasure. /	***prasāritaṃ***: expanded, swollen. ***padam kṛtvā***: to set foot in or on. ***kara***: doer, creator. ***abhya***: within the self. ***dhāraya***: holding. ***dṛḍha***: firm, strong. // **3:9**

Comment: This verse and the next two verses describe what was, no doubt, known as "girding the loins" in the ancient West. The *Haṭhapradīpikā* describes the girding as the *Mahabandha*, or Great *Bandha*.

The verse uses the word "foot" in its general sense as a "support" or "foundation." Experience teaches that an object such as a rolled cloth can be used for a source of perineal pressure if its size is adjusted not to give too much nor too little support.

3:10 Take on the form of becoming a churning vessel that produces a change in its contents by means of an inner churning rod moving up and down.

Literal: The binding in the churning vessel intensifies and holds the upper flowing energy like the sliding motion of a churning rod as a rich source of attaining generative power.

Popular: Contract the throat (in the *Jalandhara-bandha* and hold the breath in the upper part (i.e. in the *Susumna*) Then the *kuṇḍalinī* force becomes as once straight just as a (coiled) snake when struck by a rod straightens itself out like a stick.

kaṇṭhe bandham samāropya
dhārayed vāyumūrdvataḥ /
yathā daṇḍahataḥ sarpo daṇḍākāraḥ prajāyate // 3:10

kaṇṭhe: a narrow opening tube as ***kaṇṭhīla***: a churning vessel (***īla***: flow).
bandham: bind round, put on, uniting.
samāropya: transferring to, making it grow, placing in or upon.
dhārayed: holding, possessing.
vayu: wind. ***mūrdvataḥ***: ***mu***: to bind. ***urdvat***: upwards./
yathā: in such a manner as follows.
daṇḍahataḥ: struck by a churning rod. ***daṇḍa***: rod, churning stick. ***hataḥ***: forced, struck, bereft of loss, whirled up.
sarpo: slides, (moves like a snake).
daṇḍa: a rod, churning stick. ***akāraḥ***: distributes abundantly, rich source.
prajāyate: ***prajan***: procreation, generative power. ***āyat***: to attain, reachings. // **3:10**

Comment: Everyone is familiar with the inner churning sensations felt while responding to threats or demands or striving to create. Normally, the demand or desire is so strong that the body becomes like the above churn and no effort is required to put on its nature. The strong inner churning does, in fact, create a new nature of the self just as cream is converted to butter during churning. A modern endocrinologist would describe the inner churning as stimulating the production first of the tension hormones and then the production of many other hormones necessary to create the desired new form.

In the ancient writings, direct referral to the nipples is avoided. Perhaps this was due to the fears of the sensuous powers of the body that are still common today. The usage of the term "churning rod" can have no other meaning, however, than to the connection between the nipples and the *yoni* (See Chapter Nine.) This connection increases the response of both and may be necessary to initiate their sensitivity. Once the *yoni* and nipples become fully activated, the connection becomes obvious as the tensions in the *yoni* or perineum respond nearly instantaneously to the stimulation of the nipples.

Due to the neglect of the nipples, it can take considerable time to resensitize them back to their childhood sensitivity. But again, working with them as churning rods produces pleasure within a few weeks and then produces an ever-increasing response, apparently without end.

3:11 To obtain creative inner power, forcibly press down on the center of the perineum together with bending over and resting on the lower belly conquering the inner deadness.

Literal: To obtain the existence of great power, forcibly press the *kuṇḍa*. Then together with conquering the deadness of the belly by bending over to rest on.

Popular: The *kuṇḍalinī shakti* becomes straightened and the two *nadis* become lifeless.

ṛjvībhūtā tathā śaktiḥ kuṇḍalī sahasā bhavet /
tadā sā maraṇāvastā jāyate dviputāśrayā // 3:11

ṛj: to obtain, acquire. ***vibhūtā***: great power. ***tathā***: in that manner. ***śaktiḥ***: creative power. ***kuṇḍa***: hole, pitcher, bowl. ***lī***: press. ***sahasā***: forcibly. ***bhavet***: being. /	***tadā***: then. ***sā***: ***sa***: together with. ***maraṇā***: deadness. ***vastā***: ***vasti***: pelvis, the lower belly. ***jāyate***: ***jaya***: by conquering. ***dviputā***: folded double. ***śrayā***: ***śri***: to rest on. // **3:11**

Comment: At this stage of progress, the *RigVeda* needs to be consulted for the nature and production of *soma* or hormones as well as the powerful addition to the above *bandha* with the nipple stimulation produced by the ten daughters. This text states that a guru needs to be found to instruct these higher practices. This suggests that five hundred years ago, just like today, the concept of touching the nipples was politically incorrect.

Normally, the lower muscles of the belly cannot be consciously controlled, however; everyone encounters their strong contractions during vomiting, deep laughter, crying, sneezing and coughing as well as with intense emotions. At such times, the downward pressure is sometimes felt with the forced excretion of a small amount of urine out of the urethra.

This *bandha* uses the pressure gained by pressing the lower belly against the thighs such as done in the fetal pose. This pressure can serve to awaken the churning muscles similar to how pressure against the *yoni* awakens the normal sleeping *bulbosponsgiosus* muscle.

3:32 In the depths of the belly, the tongue of fire becomes active and is perceived as roaming around in the depths which indicates a change in the state of being.

Literal: In the covered cavity the tongue of fire appears to become reversed. Described as roaming around in the middle and perceived as a sign of the state of being as moving in a cavity.

Popular: To accomplish the *khechari mudra*, reverse the tongue and thrust it up the back of the throat and turn the eyes toward the eyebrows.

kapālakuhara jihva praviṣṭā viparītāgā /
bhruvorantargatā dṛṣṭirmudrā bhavati khecarī // 3:32

kapāla: cover. ***kuhara***: cavity.
jihva: tongue (of fire).
praviṣṭā: appears.
viparīta: reversed. ***gā***: obtained. /

bhru: ***bhram***: to roam around. ***antar***: in the middle. ***gat***: going.
dṛṣṭir: perceived as. ***mudrā***: a sign.
bhavati: state of being.
khecarī: movement inside the body (as a large cavity). ***khe***: in the cavity, air (***kha***: cavity, organ of generation). ***cari***: motion, travel. // **3:32**

Comment: The stimulation of the perineal area, sensed as a cavity or space in which motion can take place, initiates a sensation that feels like an inner tongue of fire as it is felt to move within the lower abdominal and perineal cavity. This experience is often found during intense worry when the effects are often described as the guts being on fire and alive or writhing deep within the belly. Similarly, during or following this experience there is often the experiencing of something flowing upward from the guts into the upper body and head. Modern physiology explains this as the flow of hormones and their capture by receptors which in turn cause changes in reactive organs.

Almost everyone is aware how the state of mind and body changes with the inner movement in the guts. For instance, a strange or threatening sound at night can induce the belly activity which in turn increases the sensitivity of the sense organs and mental awareness. If the emotion is strong enough, the emotions "take over" and control the individual to meet the impending threat.

The text is surprisingly accurate in describing the interior of the belly as a cavity since it is experienced in that manner. The Sanskrit word *khecarī*, which means "the movement through space," certainly seems applicable.

3:33 Because of the inner changes a small part of the power is loosened and carried up as high as the eyebrows.

Literal: Because of this increasing attainment it causes the increase in the loosening and separation of a small part, causing the conveyance of the power of perfection through the inner space to between the eyebrows.

Popular: This requires lengthening the tongue, by cutting the frenum of the tongue, moving and pulling it until it can reach the eyebrows.

chedanacālanadohaiḥ kalāṃ krameṇa vardhayettāvat / sā yāvadbhrūmadhyaṃ spṛśati tadā khecarī siddhiḥ // 3:33

chedana: disturbance, removing. ***cālana***: loosening. ***doha***: yielding. ***kalā***: a small part. ***krameṇa***: by means of a step, attainment. ***vardhayitr***: one who causes to increase. /	***sa***: causing. ***yāvad***: as far as. ***bhrūmadhya***: between eyebrows. ***spṛś***: to convey to. ***tadā***: then. ***khecarī***: ***khe***: air. ***carī***: moving. ***siddhi***: powers of perfection. // **3:33**

Comment: An example of this verse is given with some recent studies of laughter and crying. It is noted that when the belly is involved, tears secrete hormones which are not present at lesser levels of crying or laughing.[129] Similarly, we speak of being "blinded by emotions" often with the sense of something in the guts rising up into the head.

As the flames of *Agni* separate in rising up, an increasing inner strength, mental acuity, and physical awareness are obtained rendering the body and mind more capable of responding to outer needs or demands. The changes in the body are felt to rise up to the middle of the head where it is noted as a feeling of pressure and heat behind the eyes and a vibration or ringing in the ears.

[129] See Berk et al. (2001).

3:34 A dominating pleasure rising from the cavity is instrumental in the emission of a pungent, slippery, viscous but pure liquid. Similarly with mental guidance, sensual pleasure is obtained.

Literal: Because of a dominating pleasure, a fluid is emitted from an instrument which emits a hot, slippery, viscous but pure liquid. Similarly taking from that place in that manner mental sensual pleasure herein along with guided pleasure.

Popular: Use a sharp, lubricated knife in the shape of a cactus plant, cut the frenum a hair's breadth.

snuhīpatranibhaṃ śastraṃ
sutīkṣṇaṃ snigdhanirmalam /
samādāya tatastena romamātraṃ samucchinet // 3:34

snu: emit fluid. ***hī***: because of. ***patra***: ***pat***: to rule. ***ran***: pleasure. ***nibha***: resembling. ***śastraṃ***: tool, an instrument. ***su***: to extract or press out. ***tīkṣṇaṃ***: fiery, pungent, hot. ***snigdha***: slippery, viscous. ***nirmala***: pure. /	***samādāya***: ***sama***: similar. ***ādāya***: taking. ***tatas***: from that place. ***tena***: in that manner. ***roma***: sensual pleasure. ***mātraṃ***: ***mā***: ***manas*** mind. ***ātra***: here in. ***samucchinet***: ***sam***: along with. ***uc***: pleasure. ***chin***: mind. ***net***: ***nit***: guided. // **3:34**

Comment: This is the verse that is generally falsely translated as giving directions for starting the cutting of the frenum of the tongue a hair's breadth every day until the tongue can extend up into the sinuses.

The verse actually describes the outer secretion of a fluid from the head that is not commonly discussed but noted in times of stress, excitement or sexual stimulation. Children's noses, for instance, drain quite profusely with excitement. Adults can find drainage during tension as their eyes tear up or they "choke up" and have to clear their throats. The fluid is classified as being hot, perhaps because many times sneezing is associated with the drainage or its initiation as if you were exposed to black pepper.

Drainage is recognized to wash the eyes, the lining of the nose and sinuses, as well as being directly connected to catecholamine or adrenaline production which either is a cause or a result of it. There may also be a similar connection with serotonin and the view of the self and world and other inner neurotransmitters and hormones that respond to external threats to the individual.

Everyone is familiar with how the head "wells up" with danger or emotion, but the text describes this process as part of a total physiological and psychological response to releasing higher powers within the body. The usage of the word "wells" may be a residue of allegorizing the reaction to the falling nurturing rain or *rasana* of the god *Indra*.

3:35 After procuring and obtaining the power of *Indra*, the auspicious path can be approached and followed with repeated pounding, caressing and stroking to obtain increased collected quantity.

Literal: (No literal translation needed.)

Popular: Then rub rock salt over the tongue, after seven days, cut again.

tataḥ saindhavapathyabhyāṃ cūrṇitābhyāṃ pragharṣayet /
punaḥ saptadine prāpte
romamātraṃ samucchinet // 3:35

tataḥ: after. ***sa***: procuring. ***indha***: powers of *Indra*. ***va***: strong. ***pathya***: the auspicious path. ***abhyāṃ***: to pproach. ***cūrṇitā***: pound. ***abhyāṃ***: to approach. ***pragha***: to go forwards, proceed. ***ṣayat***: to go. /	***punaḥ***: repeated. ***sap***: caress (also sexually). ***tadin***: ***taḍit***: stroke, lightning. ***prāpte***: attained to, reached. ***roma***: ***ruh***: to increase. ***mātra***: quantity. ***samucchinet***: ***samucchi***: to collect together.// **3:35**

Comment: The inner heat and fire has been allegorized as produced by the inner god *Agni* and now the wetness is ascribed as resulting from procuring the powers of the allegorical inner god *Indra*. The early model requires that these two gods be fed with a special fluid that was called *amrita* (immortalizing fluid) similar to the Greek *ambrosia* (immortalizing fluid) also called "food for the gods." Later *amrita* became known as *soma*.

Today this process is referred to as "psyching up the body" or "stirring up the guts" in order to obtain maximum powers.

3:38 In less than half an instant of producing fluid in the head, there is a release from death and illness from food poisoning as well as illnesses resulting from growing old.

Literal: (No literal translation needed.)

Popular: The *Yogi* who remains with his tongue turned upwards even for half a *ksana* (short length of time) cannot be a victim to poisoning, disease (premature) death, or old age.

rasanāmūrdhvagāṃ kṛyvā kṣaṇārdhamapi tiṣṭhati /
viṣairvimucyate yogī vyādhimṛtyujarādibhiḥ // 3:38

rasanā: fluid. ***mūrdhva***: head. ***gā***: going. ***kṛyvā***: ***kriyā***: producing. ***kṣaṇā***: instant. ***ardham***: half. ***api***: even in. ***tiṣṭhati***: producing. /	***viṣa***: poison. ***air***: relating to refreshment or food. ***vimuc***: to be released from. ***yat***: going, moving. ***yogī***: of the *yogi*. ***vyādhi***: illness. ***mṛty***: death, disease. ***jara***: aging. ***adi***: becoming. ***bhih***: caused by. // **3:38**

Comment: The best example of this verse is given with the transformational sobbing of a child where a change occurs relatively instantaneously. Many times, crying is maintained after the transformation because it feels so good. The transformation is much faster than could be obtained from a swallowed pill and probably even faster than with a muscle injection of a drug. The recent discoveries of how hormones can increase the immune response as well as the sensitivity of the body can lend support to the protection against poisoning or age-related illnesses. The recent measured increase in response time certainly indicates that *bandha* can have a major impact upon the health and response of a body.

Rasanā is the "rain" of *Indra*, or the power of *Indra*, which is not to be confused with normal sinus drainage, since its cause is the up-flowing fire of *Agni* rather than irritation of the linings of the sinuses.

Chapter Twelve
The *RigVeda*

Introduction

The *RigVeda* (*Ṛgveda*) is a book of collected wisdom dating from over three millennia ago. The name is normally construed to mean *Hymns of Truth* but it can also be translated as *Obtaining Passion* (*ṛg*: passion and *veda*: from *vidh*: "obtaining"). I personally prefer the latter translation, since its teachings are centered upon *soma*, which is certainly the ancient name for the inner transformational hormones that were and are associated with not only creativity but passion.

The *RigVeda* not only discusses the effects of hormones but also the means for generating and controlling them. The controlling power of *soma* was called *Indra* who was described as the god of purifying rain and storms, that is, the release of passions and body fluids.

I have been amazed at the similarity of the ancient philosophy and physiology of the Greeks and the Indians, and this similarity has helped me to dig under the gloss accumulated on the writings through the millennia. Also, as I studied the *RigVeda*, I was surprised to discover that it was written for people quite similar to us today. For instance, it seems that the ancients were as squeamish about nipples and the "hidden" "secret" place in the body as are modern individuals.

Much like today, ancient societies were split into two basic groups: those who felt that all creative powers were external to an individual and a much smaller group who felt that creative power could be found within individuals.

These conclusions are evidenced by the care with which the original authors used in approaching these subjects as well as how the original works were translated in the modern academic world.

The text below gives a selection of verses of the *RigVeda* related to the control of inner hormones. The best rendition of each verse in modern English is listed first. This is followed by the 1889 academic translation by Ralph Griffith who closely followed the 1850 translation of Max Müller which was based upon the traditional interpretation of Sayana, a 14th century Hindu scholiast. The transliterated Sanskrit verse is given next and followed by a listing of the individual words and their related meanings. Finally, a commentary is given based upon a modern view. (The verses are preceded by the number of the book, the chapter, and the verse.)

Text of the *RigVeda* – Book I
"To *Agni* or the Inner Fire"

1:3:4 **Conspicuous *Indra* is the impelling force for the pressing out of *soma* (hormones) from the lower body for your needs.**

Griffith (1889): *O Indra, marvelously bright, come, these libations long for thee. Thus by fine fingers purified."*

indra yāhi citrabhāno sutā ime tvāyavaḥ /
aṇvībhis tanā pūtāsaḥ // 4

Indra: god of fluids. ***yāhiyā***: impelling. ***citra***: conspicuous. ***bhāno***: perception. ***sutā***: press out (*soma*). ***tvāyavaḥ***: for you. /	***aṇvi***: to fall to one's share. ***tan***: to be diffused. ***pūtāsaḥ***: perineum. (***puṭa***: a fold, hollow space. ***āsa***: posterior). // **4**

Comment: *Indra* was first defined as the god of rain and storms and then emotions. *Indra* is conspicuous because of the churning or pressing out of the hormones and the flow of tears.

1:3:5 Impelling *Indra* is the force for the production of *Soma* (hormones) to produce enlightenment.

Griffith: *"Urged by the holy singer, sped by song, come Indra, to the prayers, of the libation-pouring priest."*

indrā yāhi dhiyeśito viprajūtaḥ sutāvataḥ /
upa brahmāṇi vāghataḥ // 5

Indrā yāhi: impelling *Indra.*	***upa***: approach.
dhiyeśito: ***dhiya***: gifts. ***iśito***: caused.	***brahmāṇi***: one who has divine knowledge or is enlightened.
viprajūtaḥ: impelled.	***vāghataḥ***: giver. // **5**
sutāvataḥ: pressing *soma.* /	

Comment: Gut churning and the production of hormones produce creative cognition and remembrances of the brain.

1:28:1 Wherein, the broad-based lower part rises and becomes the hot creative place (or granary threshing floor) for animating *Indra*, the source of *soma* or hormones from a covered bulb.

Griffith: *"There where the broad-based stone is raised on high to press the juices out. O Indra, drink with eager thirst the droppings which the mortar shed."*

yatra grāvā pṛthubudhna
ūrdhvo bhavati sotave /
ulukhalasutānām aved v indra jalgula // 1

yatra: wherein. ***grāvā***: hard, firm. ***pṛthubudhna***: broad based lowest part. ***ūrdhvo***: to rise. ***bhavati***: becoming. ***sotave***: ***su***: bring forth, create, extract. /	***ulukhala***: ***ulu***: ***ul***: burning. ***khala***: granary, place. ***su***: ***sū***: vivify, impel, create. ***tān***: to prepare, assist. ***aved***: ***av***: animate. ***Indra***: inner creative god. ***jal***: covered. ***gula***: *lingam* or bulb. // **1**

Comment: In the modern vernacular, the above verse can be stated as tightening and raising or pulling up the seat of feelings deep in the abdomen to create an inner god-like power. The covered *lingam* is the stimulated *bulbospongiosus* muscle in the perineum (hidden and covered by the outer flesh) which is commonly experienced, when excited, as a strange urinary sensation or as a strange, warm and pleasant sexual feeling.

This verse refers to a physiological change experienced quite often during a sudden personal shock or demand often described as tightening of the anus or girding of the loins and often with excretion of a small amount of urine. A swelling (bulb) of the perineum can be noticed which includes both the *bulbospongiosus* and *pubococcygeus* muscles evidenced by the forced secretion of urine and increased sensations of pleasure and fullness.

It is difficult for most modern urban readers to identify the broad threshing floor of a granary as an allegory for their lower abdomen and the generation and purification of knowledge, vitality and mind. This model was, however, at one time popular and continues today as people refer to their lower guts as their "bread basket." The Bible (Mt 6:6), likewise refers to the lower abdomen as a lower or hidden storeroom (Greek: *tameion*) containing a source of higher powers. The *RigVeda*, unlike the extant Western writings, gives fairly detailed instructtions in how the stored power in the granary or guts can be purified and released. In ancient India, a granary was used for threshing and purifying the grain as well as for storing the grain. Oxen were tethered to a central post, and as they were driven round and round, their hoofs crushed the grain, releasing it from the husks. Greece also called the granary *kalio* which also meant a grotto containing an image of a god.

Indra symbolizes the inner fluid or hormones that allow you to change the world and your self. A modern example of the rising of the power of *Indra* would be the quiet unassuming individual who becomes the hero dashing unerringly into an unfamiliar burning house to save a child. This release of *Indra's* powers could be correlated with the gut churning and rise in hormones experienced by the hero before his dash into the house.

1:28:2 Wherein just so, the hips and the two (the hot creative place and the covered bulb), make and press out juice which vivifies and impels the release of *soma* from the *lingam* or bulb.

Griffith: *"Where, like broad hips, to hold the juice, the platters of the press are laid. O Indra, drink with eager thirst the droppings which the mortar shed."*

yatra dvāv iva jaghanādhiśavaṇyā kṛtā /
ulukhala sutanām aved indra jalgula // 2

yatra: wherein.	***ulukhala***: ***ulu***: ***ul***: burning.
dvāv: two.	***khala***: granary, place.
iva: just so.	***su***: ***sū***: vivify, impel, create.
jaghanā: pudenda, hips.	***tān***: to prepare, assist.
adhiśavaṇā: press out juice or *soma*.	***aved***: ***av***: impel.
kṛtā: produce, make. /	***Indra***: inner creative god.
	jal: covered. ***gula***: *lingam* or bulb. // **2**

Comment: Just as the feet of the oxen press out the grain from the husks on the threshing floor, so do the stimulated swollen muscles of the perineum press against the lower abdominal wall to press out an inner fluid. This fluid is now known to be hormones but was called *soma*, *amrita*, *ambrosia*, living waters or *chi* by the ancient world. The transformational powers of hormones, such as the vitalizing power of adrenaline or the cuddling power of oxytocin, were described as being like indwelling gods, and hence the inner fluids were assumed to be the food or nectar for animating the inner gods or powers. The inner gods were, therefore, explanations for superior human powers beyond those required for basic life and each separate power was described or allegorized by an anthropomorphic god.

Soma was both a god in heaven as well as an inner source of creative energy that stimulated and sustained the gods such that they could physically manifest their powers. A coined Greek word for *soma* or *ambrosia* would be *diageny* (from *dia*: "divine power" and *geno*: "creation".) "Theogony" was the Greek term to describe the birth of gods, and *diageny* can describe the birth of their powers.

In a practical sense, *soma* is the undefined and unknown source for the stimulation of organs in the body that produce various hormones and neurotransmitters within the body following the sensory or mental input of need. The varying amounts of available *soma* can explain the limited abilities individuals have to responding to external needs or demands. The *RigVeda* describes how the amount of *soma* can be increased to release the hidden inner creative powers of the body.

1:28:3 In which case, moving (the hips) like a woman forward and backward, one learns of the hot creative place which vivifies and impels an inner creative god and the covered *lingam*.

Griffith: *"There where the woman marks and learns the pestle's constant rise and fall, O Indra, drink with eager thirst the droppings which the mortar sheds."*

yatra nary apacyavam upacyavaṃ ca śikṣate /
ulukhalasutānām aved indra jalgula // 3

yatra: in which case. ***nary***: of a woman. ***apacyavam***: moving away. ***upacyavaṃ***: moving toward. ***ca***: and. ***śikṣate***: bestowing, learning to do anything. /	***ulukhala***: ***ulu***: ***ul***: burning. ***khala***: granary, place. ***su***: ***sū***: vivify, impel, create. ***tān***: to prepare, assist. ***aved***: **av**: impel. ***indra***: source of *soma.* ***jal***: covered. ***gula***: *lingam* or bulb. // **3**

Comment: This verse is perhaps the first introduction to the *Yoga* practices which are derived from the natural motions of the body to stimulate the production of *soma* or the inner heroic and pleasurable powers to meet emergencies or needs.

This motion is, of course, a sexual motion as well as an energizing dance motion. The "hot place" is, of course, that source of pleasure deep in the sexual area. The rise in activity and feelings is then related to the rising of power from that spot.

1:28:4 Wherein in the same manner, restrained churning is controlled by pressing to and fro the hot creative place which vivifies and impels an inner creative god and the covered lingam.

Griffith: *"Whereas with reins to guide a horse, they bind the churning staff with cords. O Indra, drink with eager thirst the droppings which the mortar sheds."*

yatra manthaṃ vibadhnate
raṣmīn yamitavā iva /
ulukhalasutānām aved indra jalgula // 4

yatra: wherein. ***manthaṃ***: churning. ***vibadhnate***: pressing to and fro. ***raṣmīn***: control. ***yamitava***: restrained. ***iva***: same manner. /	***ulukhalasutānām***: ***ulukhala***: ***ulu***: ***ul***: burning. ***khala***: granary, place. ***su***: ***sū***: vivify, impel, create. ***tān***: to prepare, assist. ***aved***: ***av***: impel. ***Indra***: inner creative god. ***jal***: covered. ***gula***: *lingam* or bulb. // **4**

Comment: Churning can follow the "to and fro" hip motion and is a greater stimulus to the production of *soma* or hormones. The phrase "restrained churning is controlled" is suggestive of the motion of the hips and lower abdomen during anxiety and is similar to modern statements relating to an inner uproar, binding or turmoil of the guts in response to some trauma or demand.

Unfortunately, the modern world suppresses churning by replacing it with the constant tension of a tight gut, anus, and bladder.

1:28:5 Indeed, the granary is connected to and holds an inner servant. In this place is manifested a pervading ecstasy like a victorious drum.

Griffith: *"If of a truth in every house, O Mortar, thou art set for work. Here give forth thy clearest sound, loud as the drum of conquerors."*

yac cid dhi tvaṃ gṛhegṛha ulukhalaka yujase /
iha dyumattāmaṃ vada jayatām iva dundubhīḥ // 5

yac: to ask.	***iha***: in this place.
cid: indeed.	***dyu***: manifest: ***mattā***: ecstasy.
dhi: to hold, receptacle.	***vada***: sounding.
tvaṃ: ***tva***: the place of abode.	***jayatām***: victorious.
gṛhegṛha: servant in a house.	***iva***: like.
ulukhalaka: burning inner storehouse.	***dundubhīḥ***: drum. // **5**
yujase: connected. /	

Comment: The best modern reference to the inner source of instant understanding or *gnosis* (not learned wisdom) and ecstasy is "the guts." Many people learn to trust their lower source of energy and wisdom as they "wing it" or "let it all flow out." The ecstasy, however, is generally only manifested today during vivid or REM dreams which are interpreted as being sexual-like but not sexual.

Some people remember deep, inner ecstatic feelings from childhood when they fell on their buttocks and somehow stimulated some strange inner feelings or became deliberately scared during the sharing of stories with other children. The drum is an allegory of the throbbing or pulsation that accompanies the ecstasy.

1:28:6 And always longing for and increasing audible breath to and fro. Now to find the descending powers of *Indra*, go to the inner storehouse producing *soma*.

Griffith: *"O sovran of the forest, as the wind blows soft in front of thee, Mortar for Indra press thou forth the soma juice that he may drink."*

utā sma te vanaspate vata vi vāty agram it /
atho indrāya pātave sunu somam ulukhala // 6

utā: and, also.	***atho***: now.
sma: always.	***indrāya***: powers of *Indra.*
te: they.	***pātave***: descending.
vanas: longing. ***pat***:control.	***sunu***: ***su***: to go. ***nu***: at once.
vata: audible.	***somam***: producing *soma.*
vi: to and fro.	***ulukhala***: hot granary, inner source. // **6**
vāty: to fan, wind of the body.	
agram: foremost.	
it: going towards. /	

Comment: This verse can be used to describe how some people "psyche themselves up," "turn on" or prepare themselves to meet some challenge. Children use the audible breath quite frequently in seeking inner pleasurable feelings such as a prolonged "oooooh." The term "the powers of *Indra*" would be experienced in children as the opening of the imagination or creativity as well as union with others and their world. The descending powers are later described as the rain of *Indra* which is related to tears, nasal and sinus drainage and saliva as well as the source of their stimulation.

Text of the *RigVeda* – Book Nine "*Soma Pavamāna* or Purifying *Soma.*"

9:1:1 **The bringing forth of the intoxicating purifying stream of *soma* from *Indra* is self directed.**

Griffith: *"In sweetest and most gladdening stream flow pure, O Soma, on thy way. Pressed out for Indra, for his drink."*

svādiṣṭhayā madiṣṭhaya pavasva soma dhārayā /
indrāya pātave sutaḥ // 1

svādiṣṭhayā: self-directed. ***madiṣṭhaya***: intoxicating. ***pavasva***: purifying, winnowing. ***soma***: elixir. ***dhārayā***: stream. /	***indrāya***: of *Indra*. ***pātave***: thrown. ***sutaḥ***: pressed out, brought forth. // **1**

Comment: *Soma* is probably derived from the word *su* which means to bring forth. *Soma* was first used as a synonym nickname for *amrita*. It has the same meaning as the Greek word *ambrosia* ("vital," "not dead"). The inner fluid or hormones are described as intoxicating because they overpower the normal brain thought and judgment process such as being overpowered with anger or love.

Clerics and faith healers later denied the inner production and claimed it could only be produced by their outer sanctified rituals and materials. Modern translations subvert the original concept of *soma* or *amrita* to that of an externally produced drink. This can be readily seen by comparing the literal Sanskrit with the modern translations.

The word *Indra* is the power that supplies the production of *soma* or as an allegorical god which causes its effects. It is also nearly synonymous with the powers of the inner self.[130] These comparisons are important, since the inner self is vitalized with *soma* when it becomes intoxicated, that is, not limited or subject to social rules. The bringing forth of *soma* and intoxication is accomplished by churning or winnowing of the belly or the interface between the metaphysical and physical. The term "self-directed" points to inner control rather than to external drugs or stimulation.

9:1:7 Ten subtle grasping, struggling young maidens (ten fingers) churn to produce the immediate desire, dwelling and playing within one's own place.

Griffith: *"Ten sister maids of slender form seize him within the press and hold him firmly on the final day."*

tam īṃ aṇviḥ samarya ā gṛbhṇanti yoṣaṇo daśa /
svasāraḥ parve divi // 7

tam: desire. ***īṃ*** : now. ***aṇviḥ***: ***aṇvī***: the subtle. ***samarya***: struggle. ***ā***: of course. ***gṛbh***: grasping. ***ṇanti***: to churn up. ***yoṣaṇo***: young maidens. ***daśa***: ten. /	***svasāraḥ***: one's own place. ***parve***: dwelling. ***divi***: to play. // **7**

130 See *Psuche* in Ch. 3.

Comment: To the ancient agrarian world, using the model of fingers to symbolize churning would be as clear as using fingers today for keyboarding. Five fingers or the hand might suggest masturbation, but ten fingers can only be considered as churning the contents of both breasts using the nipples as churning rods.

The aspect of playing negates following a specific regimen and rather teaches of exploring. The emphasis on one's own dwelling further emphasizes self-control and responsibility.

9:1:8 **Ten fingers deliberately incite the breasts, a gasp and a resulting thunderbolt release the invincible sweet *soma* and its three gifts (vitality, mind and intellect.)**

Griffith: *"The virgins send him forth: they blow the skin musician-like and fuse the triple foe repelling meath."*

tam īṃ hinvanty agruvo dhamanti bākuraṃ dṛtim /
tridhatu vāraṇam madhu // 8

tam: gasp for breath.
īṃ: now.
hinvanty: inciter.
agruvo: ***agruvas***: of the ten fingers.
dhamanti: ***dham***: to deliberately cause.
bākuraṃ: thunderbolt.
dṛtim: skin bag, clouds holding fluid, breasts. /
tridhatu: three parts giving.
vāraṇam: invincible.
madhu: sweet liquid (*soma*.) // **8**

Comment: As the fingers stimulate the skin bags or clouds, they cause the release of a hormone which slowly vitalizes the perineal and lowest abdominal muscles as well as neural connections for an increasing instant response. Often, there is a gasp, spasm, or jolt when the connections become fully activated which result in the initial churning in the lower belly. As churning begins, a feedback results such that the fingers move to stimulate even more pleasures within the body to further produce an increase and intensity of thunderbolts as the lower body becomes convulsed in the continuing waves of ecstasy as *soma* is produced by the churning in the belly.

9:3:10 The winnowing basket purification has many functions (such as) extracting and generating higher knowledge and power.

Griffith: *"This Lord of many Holy Laws, even at his birth engendering strength, effused flows onward in a stream."*

eṣa u sya puruvrato jajñāno janayann iṣaḥ /
dhārayā pavate sutaḥ // 10

eṣa: seeking. ***sya***: winnowing basket. ***puruvrato***: having many functions. ***ja***: produced by caused. ***jñāno***: higher knowledge. ***janayann***: generating. ***iṣaḥ***: possessing and powerful. /	***dhārayā***: possessing, preserving. ***pavate***: ***pava***: purification, winnowing grain. ***sutaḥ***: ***suta***: extracted, brought forth, generated. // **10**

Comment: This purification resulting from churning is best exemplified by the purification of mind and body resulting from the gut churning found in intense crying or laughing. This purification process was also compared to the motions used in winnowing grain in a small, handheld wicker basket which are quite similar to the motions found in churning. Somehow one discovers that churning separates out what is good from what is bad in life, similar to winnowing chaff from grain.

As one is able to keep one's heart (dedication) upon a goal, the access to the source of *gnosis* is gained and opened with winnowing. In this case, winnowing then functions to separate immediate gratifications and desires from actual needs and goals. In other words, winnowing can help you separate truth from falsehood.

9:6:1 The pleasant purifying *soma* stream pouring forth devoted to gods animates the attainment of goals.

Griffith: *"Soma, flow on with pleasant stream, a bull devoted to the gods, our friend, unto the woolen sieve."*

mandrayā soma dhārayā vṛśā pavasva devayuḥ /
avyo vāreśv asmayuḥ // 1

mandrayā: pleasant, hollow rumbling vibrations. ***soma***: elixir. ***dhārayā***: flowing stream. ***vṛśā***: pour forth, effuse. ***pavasva***: purification. ***devayuḥ***: devoted to gods. /	***avyo***: ***av***: to lead or bring to. ***āvyā***: to drive, impel, animate. ***āvī***: approach, to grasp, seize. ***vāreśv***: able to grant wishes. ***asmayuḥ***: desire. // **1**

Comment: This verse suggests that churning-winnowing directed to yielding a particular power expedites the inner creational process. A simple model is that in facing a physical threat, the seeking for the power of a famous warrior is far more advantageous than some general desire such as becoming successful. Many modern individuals would describe this process as taking on the nature of some god or hero or allowing that god to take over control of the situation and the self. This process assumes, of course, that one has already activated the process of generating *soma.*

It might be of interest to note that the phrase *avyo vāreśv* is commonly translated as "woolen filter" in support of an assumed external process used to purify a psychedelic plant juice instead of accepting the existence of an inner produced *soma* (or hormone) which has its own powers.

9:6:5 **Indeed the extraordinary impetuous rubbing ten (fingers) acquire, raise, and bestow great animation.**

Griffith: *"Whom having passed the filter, ten dames cleanse, as 'twere a vigorous steed. While he disports him in the wood."*

yam atyam iva vājinam mṛjanti yośaṇo daṣa /
vane krīḷantam atyavim // 5

yam: confer.
atyam: ***ati***: extraordinary, very great.
iva: indeed, just so.
vājinam: impetuous.
mṛjanti: wiping, rubbing, purifying.
yośaṇo daṣo: fingers. /
vane: acquire.
krīḷantam: ***kṛ***: fill with, purify, cause, raise up, place. ***īḷa***: animation, vital spirit.
atyavim: ***atya***: great. ***vim***: to bestow. // **5**

Comment: Those who discover methods of stimulating the nipples or breasts generally find subtle methods of self-stimulation to increase their vitality or as a method to raise themselves from boredom. Men are able to directly rub their breasts but they are also able to bring their hands to their suspenders or upper vests with subtle rubbing or applying pressure to the nipples. Women are socially forbidden to directly rub their breasts but learn to use such gestures as fluttering their hands or arms to lightly stroke their breasts.

9:14:3 **Mighty *rasa* (*soma*) exists and imparts heavenly bliss to all the gods, hence, contributing to the radiance of enlightened people.**

Griffith: *"Then in his juice whose strength is great, have all the gods rejoiced themselves, when he hath clothed him in the milk."*

ād asya ṣuśmiṇo rase viṣve devā amatsata /
yadī gobhir vasāyate // 3

ād: ***dā***: impart. ***asya***: to be, exist. ***ṣuśmiṇo***: ***śuṣ min***: fiery, strong. ***rase***: ***rasa***: essential juice of the body, nectar. ***viṣve devā***: all the gods. ***amatsata***: to enjoy heavenly bliss. /	***yadī***: hence. ***gobhir***: ***go***: star, light, cow (enlightened). ***bhir***: people. ***vasā***: radiance, shining. ***yate***: to strive to obtain anything, extends. // **3**

Comment: *Rasa* or *soma* is described as the power which feeds the gods or the source of powers required to manifest a *mantrā* or vision. This *rasa* can be interpreted today as the power to correctly select and determine quantities of hormones. There is the strong implication that if the gods or powers are forced with the effort of the brain and body, problems will be encountered. It must also be reiterated that although gods are but allegories of the inner powers of individuals, they are very useful in defining and specifying that power.

When the gods are filled with *soma* and bliss, they can be perceived as having a radiance commonly noted with the manifested powers of creative or enlightened individuals.

9:14:4 Without a living master, one prepares and quickly moves along another path to prepare the self at this time to unite, conquer and destroy.

Griffith: *"Freeing himself he flows away, leaving his body's severed limbs, and meets his own companion here."*

niriṇāno vi dhāvati jahac charyāni tanvā /
atrā sam jighnate yujā // 4

niriṇāno: ***nir***: without. ***ina***: master. ***ana***: living. ***vi***: go another direction. ***dhāvati***: quick. ***jahac***: ***ja***: produced by. ***hā***: to fall or come into any state. ***charyāni***: ***car***: moving. ***yāni***: a path. ***tanvā***: ***tanu***: self, body. ***tan***:to extend, accomplish, prepare a way. /	***atrā***: at this time. ***sam***: together with. ***jighnate***: ***ji***: to conquer. ***ghna***: destroy. ***yujā***: to unite. // **4**

Comment: The living master is the active control by the socially conditioned brain which the individual has been taught to obey which includes the constant judgment as to how he is behaving or thinking. The other direction is breaking free of the demand of social conditioning in order to set and obey what is within your own heart or dedication in life and beyond. The *RigVeda* conceives of God as having three attributes: creative, destructive and maintaining which must also be attained by an individual to change his or her own world. Many of the ancient religions taught that individuals had to become as gods in order to become enlightened.

9:14:5 The fingers move to bring together the purity of youth and an inner fire to purify the self.

Griffith: *"He by the daughters of the priest, like a fair youth, hath been adorned, making the milk, as 'twere, his robe."*

naptībhir yo vivasvataḥ subhro na māmṛje yuvā /
gāḥ kṛṇvāno na nirṇijam // 5

naptībhir: ***napati***: daughters, hand and fingers. ***bhir***: by means of. ***yo***: bring together. ***vivasvataḥ***: shining forth, inner fire. ***subhro***: ***ṣubhra***: radiant. ***na***: not. ***māmṛje***: ***mā***: not. ***āmṛje***: ***yuvā***: youth. /	***gaḥ***: pursue, moving. ***kṛṇvāno***: obtain. ***na***: not. ***nirṇijam***: ***nir***: without. ***nij***: purify one's self, clean. // **5**

Comment: Becoming like a god requires an inner sense of certainty and purity that children have in abundance. This purification can perhaps best be suggested by remembering how clean and pure the world and self appeared after an intense crying or laughing spell.

It is discovered that the impetuous fingers on the breasts automatically magnify an intense and deep feeling which accomplishes the cleansing even better than crying or laughing. The inner fire results from the rise in eagerness to change the self and world.

9:14:6 The fingers cause the transverse attainment of an enveloping enlightenment with an extensive vibration which expands knowledge.

Griffith: *"O'er the fine fingers, through desire of milk, in winding course he goes, and utters voice which he hath found."*

ati ṣritī tiraṣcatā gavyā jigāty aṇvyā /
vagnum iyarti yaṃ vide // 6

atiṣritī: to cause to pass through. ***tiraṣcatā***: transversely. ***gavyā***: ***ga***: enlightenment. ***vya***: to envelope one's self. ***jigāty***: ***gam***: to attain any state. ***aṇvyā***: ***aṇu***: ***aṇvi***: by means of fingers. /	***vagnum***: vibration, sound. ***iyarti***: ***iyat***: of such extent. ***yaṃ***: ***ya***: expands. ***vide***: knowledge. // **6**

Comment: The modern individual looks for enlightenment as being something like a beam of light coming down from the heavens. Instead, the whole world and the inner self changes transversely rather than vertically. Enlightenment requires that your entire world contains light rather than experiencing some isolated good experience or understanding.

The extensive vibration is commonly experienced as a feeling of beauty, joy, perfection, or power existing in your world. Children generally experience the feeling being accompanied with a penetrating high pitch sound called the *nādam* by the ancients and tinnitus by science. The sound can be heard by some adults during a body or mental stress or during excitement. The vibration can be described as being associated with an energy flow which changes the self and world.

9:14:7 Moving rapidly towards conflict and uniting with it as the possessor of powerful purification, stand forth grasping and bowing to the heroic.

Griffith: *"The nimble fingers have approached, adorning him the Lord of Strength: They grasp the vigorous courser's back."*

abhi kśipaḥ sam agmata marjayantīr iśas patim /
pṛṣṭha gṛbhṇata vājinaḥ // 7

abhi: towards.
kśipaḥ: moving hastily.
sam: uniting with.
agmata: ***agman***: battle, conflict.
marja: purification.
yantīr: holding.
iśas: powerful.
patim: master, possessor. /
pṛṣṭha: prominent, standing forth.
gṛbh: grasping, seizing.
ṇata: bowing to.
vājinaḥ: ***vājin***: heroic. // 7

Comment: With the return of the inner powers known to children comes the same strong desire to face and conquer the outer world. The *RigVeda* wisely equates the purification found in facing the challenges of life as the greatest source of purification for an individual.

However, the state of enlightenment must be first reached to fully profit from facing life. Honoring the heroic constitutes a dedication that overcomes the conditioned mode to be a helpless subject to life relying upon others to solve problems.

9:15:1 This source of *gnosis* and the hidden inner powers are entirely produced by the ten maidens or fingers, pleasure, vitality and (the resulting) flow of *soma* or hormones.

Griffith: *"Through the fine fingers, with the song, this hero comes with rapid ears, going to Indra's special place."*

eśa dhiya yaty aṇvya ṣuro rathebhir aṣubhiḥ /
gachann indrasya niśkṛtam // 1

eśa: ***etad***: this.
dhiya: *gnosis*, divine knowledge.
yaty: ***yati***: giver, source.
aṇvya: ***aṇvi***: by fingers preparing *soma*.
ṣuro: flow of *soma*.
rathebhir: by pleasure.
aṣubhiḥ: by. ***aṣu***: vitality, spiritual life. /
gachann: ***ga***: moving.
channa: hidden.
indrasya: of the inner powers.
niś: entirely. ***kṛtam***: created, gained, accomplish, encompass. // **1**

Comment: At first reading, this verse is contrary to popular belief. Instead of arguing that effort and desire for powers bring forth those powers, it states that it is pleasure which gives rise to the higher powers evidenced by great people.

Children present the best evidence of its truth. Children vitalize themselves or find pleasure in being shocked, scared or in placing new demands upon the body or imagination. Children are also able to find vitality and hence special powers in becoming sensual and responsive to real and imagined forces affecting the body such as enchantments and bondage. The stimulation of ten "impetuous" fingers is readily observed in "tickling" games as well as in their stimulation found in preparing to sleep, while dreaming, and during mental and physical pain.

9:36:2 Procure *soma*, the overflowing sweetness from the treasury, which awakens, purifies, gratifies, and feeds the gods or inner powers.

Griffith: *"Thus soma, watchful, bearing well, cheering the gods, flow past the sieve, turned to the vat that drops with meath."*

sa vahniḥ soma jāgṛviḥ pavasva devavīr ati /
abhi kośam madhuścutam // 2

sa: procure.
vahniḥ: *soma*, the conveyer or bearer of oblations to the gods.
soma: elixir.
jāgṛviḥ: awake.
pavasva: purify.
devavīr: gratifying the gods.
ati: to pass on. /
abhi: towards.
kośam: treasury, cloud.
madhuścutam: overflowing sweetness, verpowering intoxicating drink. // **2**

Comment: This verse reflects the ancient view that the lower heart or center of the lower abdomen stores up the source of energy or vitality. Their comparison of the source of *soma* to a cloud is excellent, since they were aware that the source was scattered much like a cloud and can be directly compared with the production of hormones from the intestines.

9:93:1 The ten sisters (fingers) moving quickly together with unrestrained rubbing emit and bestow pleasure. Produced by traversing and bringing forth from one's chest like a slow-moving cloud pouring forth its contents to bring forth extraordinary strength.

Griffith: *"Ten sisters pouring out the rain together, swift moving thinkers of the sage adorn him. Hither hath run the gold-hued Child of Sūrya and reached the vat like a fleet vigorous courser."*

sākamukśo marjayanta svasāro daśa
dhirasya dhītayo dhanutrīḥ /
hariḥ pary adravaj jāḥ sūrasya droṇaṃ nanakśe atyo na vāji // 1

sāka: together. ***ukśo***: emit. ***marj***: rubbing, to play as a drum, cleaning. ***yantur***: not regulated. ***svasāro***: sister, one's own body. ***daśa***: ten. ***dhirasya***: ***dhi***: to satisfy, delight, hold. ***rasya***: pleasure. ***dhītayo***: ***dhīta***: bestow. ***dhanutrīḥ***: ***dhanutṛ***: moving quickly. /	***hariḥ***: carrying. ***pary***: around. ***adravaj***: ***adrava***: not running, swift, liquid. ***jāḥ***: produced. ***sūrasya***: ***su***: to bring forth. ***urasya***: coming from the chest. ***droṇaṃ***: a cloud (carrying fluid), bucket, *soma* vessel. ***nanakśe***: ***na*** (1): ***nakś***: obtain. ***atyo***: exceeding. ***na*** (2): most certainly. ***vāji***: strength, vigor. // **1**

Comment: This verse describes the feelings that are finally obtained once the nipple stimulation is mastered. As the chest and nipples become sensitive with the development of ecstatic feelings, they can easily be imagined to produce a fluid like rain dropping from heaven to the earth or perineum below where it activates the soil or the ground.

Sanskrit uses references of rain and lightning to imply the powers of *Indra* who is the god of rain and lightning as well as light. The breasts are also described as clouds holding the rain of *Indra* for its release to the *yoni*, the middle of the perineum, below. This is an argument for why the mammae of humans differ from that of animals.

9:93:2 **Together (the fingers) cause the residence of the source to swell without fear producing a flow very difficult to be restrained or separated. There are no limits to entering the light with the fingers carnally approaching and extracting from the breasts.**

Griffith: *"Even as a youngling crying to his mothers, the bounteous Steer hath flowed along to waters. As youth to damsel, so with milk he hastens on to the chose meeting-place, the beaker."*

sam mātṛbhir na śiśur vāvaśāno
vṛśā dadhanve puruvāro adbhiḥ /
maryo na yoṣām abhi niṣkṛtaṃ
yan saṃ gachate kalaśa usriyābhiḥ // 2

sam: together.
mātṛbhir: ***mātṛ***: source.
 bhir: ***bhi***: to separate.
na śiśur: ***śi***: to cause.
 śū: ***śvi***: to swell.
vāvaśāno: to remain, produce, cause to exist.
vṛśā: ***vṛś***: the fingers.
da: producing.
 dhanve: to flow.
puruvāro: ***puru***: much.
maryo: ***maryā***: limit.
na: no, not.
yoṣām: ***yoṣan***: the fingers.
abhi: to, towards.
niṣkṛtaṃ: to extract.
yan: ***yād***: to be closely united or connected with.
saṃ: together.
gachate: ***gam***: to approach carnally, to go to any state, going.
kalaśa: breasts, cloud.

vāra: difficult to be restrained. ***adbhiḥ***: ***ad***: consume. ***bhiḥ***: ***bhi***: to separate. /	***usriyābhiḥ***: ***usriya***: light. ***abhi***: to enter. // **2**

Comment: The stimulation is from the breasts. The residence of the source of power is in the sexual region or the belly and when activated is associated with swelling of the perineal muscles and tissues.

The phrase "without fear" suggests that the ancients also had strong feelings against stimulating the body which might be similar to the modern fear of massaging nipples.

The usage of the word "carnal" is important, since touching and being touched both increase seemingly without limit, which is suggestive of sexual arousal except that the breast stimulation is not immediate, but rather increases at a steady rate little by little. The pleasure is unique in that after it is discovered it is perceived as being at the limit of tolerance yet the level of tolerance keeps increasing as does the intensity of pleasure. This is, of course, opposite to most experiences of sexual pleasure which quickly reach a peak during each experience with little of no increased pleasure after repeated efforts. Like the sexual response, and because of the production time of hormones, there is a delay after the initial stimulation of the nipples and the beginning of the lower purifying churning.

9:93:3 **Also, great swelling of the breast with the power of *Indra* like a cloud contains the available nourishment for the mixing and churning in the beneficent highest purifying vessel.**

Griffith: *"Ye, swollen is the udder of the milch-cow: thither in streams goes very sapient Indu. The kine make ready, as with new-washed treasures, the head and chief with milk within the vessels."*

uta pra pipya ūdhar aghnyāyā indur
dhārābhiḥ sacate sumedhāḥ /
mūrdhānaṃ gāvaḥ payasā camūśv
abhi ṣrīṇanti vasubhir na niktaiḥ // 3

uta: also.	***mūrdhānaṃ***: ***mūrdha***: the highest.
pra: great.	***gāvaḥ***: spiritual.
pipya: to swell.	***payasā***: fluid.
ūdhar: ***ūdhas***: breast.	***camūśv***: vessel.
aghnyāyā: being like a cloud.	***abhi śri***: to mix. ***ṇanti***: ***mantha***: churning.
indur: *Indra.*	***vasubhir***: ***vasu***: excellent, good, beneficent. ***bhir***: by means of.
dhārābhiḥ: ***dhāra***: containing. ***abhi***: to, towards.	***na***: like.
sacate: available.	***niktaiḥ***: ***nikta***: purified. // **3**
sumedhāḥ: nourishing. /	

Comment: Swelling of the nipple and its underlying structure is known to occur with repeated stimulation. Men's nipples are able to at least double their size as they develop sensitivity and become capable of stimulating the lower abdominal and sexual muscles. Since the effects of the stimulation lasts for many hours even after the stimulation of the nipples has ceased, the effect has been allegorized as being due to the charging of the chest with a fluid similar to that of clouds which then slowly drop the fluid similar to clouds losing their contained water to the ground below.

The released fluid or hormones of the chest is able to stimulate the pulsation of the lower muscles which are unable to be controlled except during such body requirements found in deep vomiting, sexual orgasm, or prolonged, intense sobbing or laughter.

The highest or chief purifying vessel is often symbolized as a winnowing basket in Indian writings as well as in early Greek because of its physical motion as well as its ability to purify and change one's world.

RigVeda Text – Book Ten
"To *Agni* or the Inner Power"

10:85:2 (Because of) *Soma's* ubiquity and *soma's* physical power, *soma* is therefore non-ceasing, non-moving and secure in the small dwelling place of the gods.

Griffith: *"By soma are the Adityas secure. By soma mighty is the earth. Thus soma in the midst of all these constellations hath his place."*

somenāditya balinaḥ somena pṛthivi mahi /
atho nakśatrāṇām aśam upasthe soma ahitaḥ // 2

somena: belonging to *soma*. ***aditya***: ***aditi***: boundless.	***atho***: also, now.
balinaḥ: powerful.	***nakśatrāṇām***: abode of gods. ***nakśatrā***: heavenly body. ***aṇā***: small.
somena: belonging to *soma*.	***a***: not. ***śam***: extinguished.
pṛthivi: earth.	***upasthe***: secure place.
mahi: physical earth. /	***soma ahitaḥ***: ***a***: not. ***hita***: moving. //2

Comment: This verse argues that because *soma* is a cosmic power it is also present within an individual and supplies its power to the inner powers or gods of an individual as well as to the gods or powers of the expansive heaven.

10:85:3 ***Soma* passing as an elixir physically prepared from plants is not the *soma* described by the wise sage, and not even that used for oblations or anointing.**

Griffith: *"One thinks, when they have brayed the plant, that he hath drunk the soma's juice; of him whom the Brahmans truly know as soma no one ever tastes."*

somam manyate papivan
yat sampiṅśanty oṣadhim /
somaṃ yam brahmaṇo
vidur na tasyaṣnāṣti kaṣ cana // 3

soma: ***manyate***: passing for. ***papivan***: elixir, drink. ***yat***: prepared. ***sam***: with. ***piṅśanty***: ***pinaṣṭi***: grinding with instruments. ***oṣadhim***: herb, plant. /	***somaṃ***: *soma*. ***yam***: pronounce, displayed, offered. ***brahmaṇo***: sage. ***vidur***: ***vidu***: wise. ***viduh***: exploiting. ***na***: not. ***tasya***: his. ***ṣnā***: ***snati***:oblations. ***kaṣ***: rub. ***cana***: certainly not. // **3**

Comment: This verse is quoted by individuals who attempt to argue that *soma* is not obtained from plants through complex religious procedures and rituals by clerics or priests. It is noteworthy, however, that people even 4,000 years ago preferred to believe in external cures and controls rather than assuming responsibility and effort for their own perfection or evolution.

Chapter Thirteen
The *Rudrayāmala*[131]

Introduction

This document has been politically incorrect for not just the last fifty years but for more than two millennia. Its central theme is finding an identity with the inner Creator God *Brahmā.*[132] Those people who have found this inner God have been called by such names as the enlightened, righteous, and liberated. When individuals become enlightened or righteous, they do not need religion and instead can find truth from their own heart (between the thighs).[133] The enlightened have an active role in the world but are not in any bondage to it.

The *Rudrayāmala* is not a "how-to manual" for the unrighteous but rather a description of the powers that produce righteousness. Society does not acknowledge the existence of the righteous much less offer an explanation of how they are able to do the things they do.[134] The righteous seldom find a description or even an acknowledgement of what makes them different. This writing is therefore important to those approaching enlightenment since it describes what happens during the process of opening the body and mind that is actually denied by most modern "developed" societies.

The original *Rudrayāmala*, which is at least as old as the *RigVeda*, has been lost, but its content has survived,

---

[131] *Rudra*: "Creative Power," *yā*: "attaining," *amala*: "purity"

[132] The personal inner power is called *Brahmā* indicating the relationship to heavenly *Brahma.*

[133] Matthew 9:12

[134] See Chapter 8.

embedded within writings such as the ancient *Parā-triṃśikā*,[135] which I found embedded in the *Parātrīśikā Vivaraṇa.*[136] This Sanskrit text was written by the eleventh century *Tantrik* scholar known as Abhinavagupta.[137] Surprisingly, Abhinavagupta did not explain the meaning of the contained *Rudrayāmala* but did state how the meaning could be extracted from the ancient Sanskrit. His dissertation provided me with an insight into how Sanskrit documents could have meanings intentionally hidden from people who were not ready for the teachings. Abinavagupta describes how ancient Sanskrit hides teachings by combining descriptive words together to form a single word with another meaning using *sandhi* rules.

The hidden *Rudrayāmala* requires the understanding of critical terms. If these terms are understood, and if the *sandhi* connected words are separated correctly, a literal translation can result which utilizes and agrees with every word. Hence, the correctness of a translation can be proven millennia later.

Rudra, one of the oldest gods of India, was described as coming forth from *Brahma*,[138] in the same manner as *Eros* came forth from *Zeus* in Greek legends. Both *Rudra* and *Eros* are names for the source of energy which gave reality to the metaphysical creative visions of *Brahma* and *Zeus*. *Rudra* has heavenly attributes as well as inner individual powers as does *Eros*.

---

[135] Translates into English as "Thirty Verses on the Supreme."

[136] *vivrana*: "an expansion on"

[137] *abhi*: "bringing forth," *nava*: "modern (interpretations)," *gupta*: "secrets"

[138] self-existent, absolute, eternal.

Similarly, *Rudra*, like *Eros*, has both a creative and an inner destructive power. The creative powers of *Rudra* are much better defined than those of *Eros* and are credited to being carried out by subordinate gods (the *Maruts* or *Rudras*) such as *Iśvara* for the power of Love, *Nir ṛita* for the power of destruction, *Mṛta* for the power of vitality and action, and *Sarpa* for the visionary power of the snake or serpent.[139] *Rudra* also has the support of *Agni*, the god of fire, and *Indra*, the god of rain. *Agni* can supply the inner heating flames associated with sobbing or other emotions, while *Indra* supplies the free flowing and cleansing fluids from the eyes, nose, sinuses and mouth.

Naming *Rudra's* inner creative powers as separate gods or powers allowed the ancients to compare them to the effects of hormones as well as to describe practices that increase or control individual powers. An excellent example of this selective control is the ability of actors and actresses to change their inner hormones to project different feelings such as increasing the flow of *Indra's* rain as tears or increasing the inner fires of *Agni* for the inner intensity of emotions. The *Rudrayāmala* explains that the individual powers, once activated, are controlled by the heart by "putting on a role." This consists of mentally creating a new *mantra*[140] or vision of what is desired and then manifesting it as a physical reality or *mudra* through the integration of the inner powers or hormones.

One startling characteristic of the power or god *Rudra*, which I discovered is quite important in understanding the following document, is that *Rudra* (or *Śiva*, the later

---

[139] The snake, serpent and dragon used to be the allegory for positive forces.

[140] An instrument of thought used to envision a new role, followed by the manifesting of it as a *mudra*.

name of *Rudra*) has both male and female attributes.[141] *Rudra*, representing inner powers, is both male and female as commonly described but also has the higher characteristics that both sexes can develop. *Rudra* can be perceived as a god or an ideal state of having fully developed the *yoni* and its response in the middle of the perineum. The *yoni* is synonymous with the outer control and opening of the sacral heart and is depicted as a pudendum with a protrusion of flesh[142] as described in the *RigVeda* and the *Haṭhapradīpikā*.[143]

The *Rudrayāmala* has become a much disparaged text because it is not religious and describes what is now considered to be a prurient or suspiciously sensuous method of generating and controlling *soma* or what modern science would call the inner transformational hormones. The *Rudrayāmala* describes how the higher powers, such as described by Maslow, are attained. These powers include the ability to touch a greater source of knowledge and freedom than commonly found in our society. The methods described in the text are seemingly abhorrent. That this abhorrence is not based upon any reasonable logic should illustrate to the reader how the modern brain has been socially conditioned.

Each verse begins with a twenty-first century translation followed by a literal translation. Next, the transliterated original and separated Sanskrit words and roots are included along with their dictionary meanings. The comments will attempt to explain the technical content of complex verses in modern terms.

---

[141] These attributes are also associated with *Eros*.

[142] See Ch. 9.

[143] See Ch. 11 and 12.

Text of the *Rudrayāmala*

1. I ask myself, how do I find the power of discernment within my mind and the power in the churning of my guts that opens me to being enlightened?

Literal: The Soul asks the inner God of one's own self how the higher powers are obtained and conferred by the sexual region by means of discernment and churning to find the higher path.

śrī devī uvāca /
anuttaraṃ kathaṃ deva svataḥ kaulikasiddhidam /
yena vijñātamātreṇa khecarī-samatāṃ vrajet // 1

śrī devī uvāca: the Soul speaks. /	***yena***: by means of.
anuttara: chief, principle.	***vijña***: discern, know, understand.
katha: what, how.	***mātr***: ***mathra***: to churn.
deva: inner mind or god.	***eṇa***: by means of.
svataḥ: of one's own self.	***atra***: in this place.
kaulika: ***kaupina***: sexual region.	***eṇa***: by means of.
siddhi: higher powers.	***khecarī***: aerial or higher being, sky path.
da: conferring. /	***samatāṃ***: identify with.
	vraj: obtain. //**1**

Comment: "Asking one's self " is a step beyond the normal rumination of the brain and awakens the inner Soul.[144] This inner Soul or Self can only be described as the power behind the gut instincts or the self-activating power that arises spontaneously when a crisis is faced, but this inner, intuitive Self is not the same as the normal judgmental self who worries about conformity, etc. This Soul or *Devī* was known as the inner feminine goddess *Aphrodite* in Greece whose yearning for union

[144] See Ch. 3.

and perfection gave birth to the inner masculine god *Eros* (known herein as *Deva*, *Bhairava*, *Brahmā* or *Rudra*) who could manifest her dictates.

Understanding of the nature of the Soul was gradually lost in both the West and the East as the Soul became redefined as the socially conditioned self-identity which obeys social law. In modern India the mentally identified self is now known as the *atman* (See Verse 32) and in the West it is known as the psyche. Just as the knowledge of the Soul has faded, the awareness of an inner power that could manifest the Will of the Soul has been nearly lost as *Eros* became Cupid in the West and *Rudra* became *Śiva* in India.

The verse also agrees with the early Greek model of ascribing the nature of different gods to the various inner powers of an individual.

The churning of the guts is experienced frequently when facing a challenge and is now recognized as stimulating the production of hormones able to assist in overcoming external problems. The modern problem with churning is that it is largely suppressed and the supporting muscles weakened though disuse.

2. **Within my sacral heart exists the connection between the power in the sexual or gut region and the Soul.**

Literal: The hidden, sexual region is my inner counsel and light. Existing in the heart is the strong connection of the power of the region and the mistress.

etad guhyaṃ mahaguhyaṃ
kathayasva mama prabho /
hṛdayasthā tu yā śaktiḥ kaulikī kulanāyikā // 2

etad: this. ***guhyaṃ***: hidden, secret pudendum. ***mahaguhyaṃ***: very hidden. ***kathayasva***: converse with. ***mama***: mine. ***prabho***: ***prabha***: powerful, leader, light. /	***hṛdayasthā***: existing in the heart. ***tu***: strong. ***yā***: going. ***śaktiḥ***: connecting power. ***kaulikī***: ***kaulina***: sexual region. ***kula***: abode. ***nāyikā***: mistress. // **2**

Comment: The usage of the word "heart" as the dwelling place for the Soul has become distorted in modern India as well as the West. The heart is now considered to be in the blood pumping organ in the chest which follows if the Soul is falsely redefined and limited to be only the force of vitality. The lower true heart[145] in the sacrum is the interface between the metaphysical Soul and the physical inner power known as *Eros* or *Rudra*.

3. The Soul asks: tell me *Rudra*, how do I become enlightened? And he replies, by the Soul envisioning the future desires of the heart.

Literal: Tell me chief God (*Deveśa*), by what means do I find fulfillment? *Rudra* states, the illustrious Soul (Devī) great dispenser of the selected delightful future to be reached.

tāṃ me kathaya deveśa yena tṛptiṃ labhāmyaham /
śri bhairava uvaca śṛṇu devi mahabhage
uttarasyāpyanuttaram // 3

tām me kathaya: tell me how. ***deveśa***: of chief god,	***śri bhairava uvaca***: *Rudra* speaks. ***śṛṇu***: ***śṛī***: illustrious.

[145] See Ch. 1.

Brahmā. ***yena***: by which means. ***tṛpti***: fulfillment. ***labhāmy***: find. ***aham***: I. /	***ṇu***: ***mu***: praise. ***devi***: ***devī***: goddess. ***mahābhāge***: great dispenser. ***uttarasyā***: belonging to the future. ***āpya***: to be reached. ***nutta***: sent, ordered. ***ram***: to delight in. // **3**

Comment: This verse states clearly that the Soul cannot assume that the inner powers will bring fulfillment. It must come by the Soul reaching ever further into the future. Instead of expecting some inner power to produce enlightenment, the Soul must assume responsibility.

4. My conscious brain asks my Soul in my awakened heart, what path must I follow to find optimum relationships with the outer world?

Literal. Possessing a ruling goddess in my heart existing in my perineum, what path, goddess of gods, must be mastered to bring evolutionary powers to that area?

**kauliko'yaṃ vidhirdevi mama hṛdvyomnyavasthitaḥ /
kathayāmi sureśāni sadyaḥ kaulikasiddhidam // 4**

kaulikī: ***kaulina***: sexual region, perineum. ***ayam***: to possess. ***vidhi***: worshipper, rule. ***devi***: goddess. ***mama***: mine. ***hṛd***: heart. ***vyom***: heaven. ***nava***: new. ***sthitaḥ***: standing. /	***katha***: what. ***yāmi***: path. ***sureśa***: goddess of gods. ***ani***: to bring, use. ***sadyaḥ***: to be subdued or mastered or won or managed. ***kaulika***: ***kaupina***: private part. ***siddhidam***: conferring evolutionary power. // 4

Comment: Individuals must first open their hearts to the outer world before setting their own paths to evolve.

5. **The inner power provides protection from sin as well as integrating the physical and metaphysical worlds.**

Literal: Now, this existing strong protector of your entire world obtains shining freedom from sin, and that inner part manifests the subtle connection of the Sun and Moon.

athādyāstithayaḥ sarve svarā vindvavasānagāḥ /
tadantaḥ kalayogena somasūryau prakīrtitau // 5

atha: now. ***adya***: this. ***asti***: existence. ***tha***: protector. ***ya***: strong. ***sarve***: all. ***svar***: world. ***vinda***: acquires. ***vasa***: shining. ***anāgās***: sinless. /	***tad***: that. ***anta***: inner part. ***kalayogena***: subtle connection. ***somasūrya***: sun-moon. ***prakirt***: to manifest. // **5**

Comment: Enlightenment requires the ability to connect the metaphysical and physical forces together such as creativity and manifesting, dreaming and reality, eternity and life etc. When one creates and manifests one's own world there can be no sin since everything must fit together. This was perhaps the greatest surprise of Maslow when he studied the Great People and found that they could essentially do no wrong.

The Soul interfaces with a higher metaphysical controlling power known by many names such as *Zeus* to the Greeks and *Brahma*[146] to the Indians. In this mod-

[146] The Self-Existent; compare with Yehovah. Note the personal

ern age it is easy to simply consider the higher controlling power as being in the form of Natural Law which serves to direct, limit and control energy-matter-intelligence interactions.

6. (The Soul) becomes the source of filling an individual with truth, pleasure, as well as a sense of purpose and immortality. An expanding evolving life is then met step-by-step, limited only by an inner image of perfection.

Literal: Bestowing, covering and infusing the self with truth, pleasure, and noble purpose. Approaching constantly step-by-step and expanding a created state with virtuous boundaries.

pṛthivyādīni tattvānī puruṣāntani pancasu /
kramātkādihu vargeṣu makārānteṣu suvrate // 6

***prithi*: *priti*: pleasure. *vyā*: a coverer. *ādīni*: noble minded. *tattva*: truth. *ani*: to bring to. *puruṣa*: animating principle, spirit. *antani*: within. *panca*: spread out. *su*: to bestow. /**	***kramā*: approach by steps. *atka*: *at*: go constantly. *dih*: increase. *varga*: state. *ṣu*: *sū*: to bestow. *ma*: *maṃh*: to bestow. *kāra*: create. *anta*: limit, boundary. *ṣu*: *su*: to bestow. *suvrata*: virtuous. // 6**

Comment: This verse elaborates upon the verse above and explains that the union with the superior Higher Power is pleasurable. The intent of the self to find perfection and beauty can be considered as opposite to the normal wasting away and depletion of the assets of the earth. A living individual is able to direct the reversal of

God is *Brahmā* not *Brahma*.

entropy or the loss of energy and order. The energy for that reversal is supplied by the mysterious Life Force which is present in all forms of life.

The remainder of the *Rudrāyamala* describes how this change can be accomplished as well as the results of that change.

7. **The four basic elements of the world, earth, air, fire and water, contain their own inner power to change as required in an evolving world.**

Literal: There are four supporting elements of reality: air, fire, flowing water, and manifested earth. From these there arises a shining forth preceding the developing of the expansive world.

vāyvagni salilendrāṇām dhāraṇānam catuṣṭayam /
tadūrdhvaṃ śādi vikhyataṃ
purastāt brahmapancakam // 7

vāyvagni: ***vayu*** and ***agni***: air and fire.	***tad***: that. ***ūrdhvaṃ***: rising.
salilendrāṇām: *Varuna's* city: water.	***śādi***: eminence.
dhāraṇānam: maintaining earth.	***vikhyataṃ***: beholding, shining.
catuṣṭayam: fourfold. /	***purastāt***: before.
	brahmapancakam: fifth creation, evolutionary. // **7**

Comment: Air, fire, water, and earth are universal mystical entities with metaphysical properties that survive even today as defining the foundation of the physical world as *length*, *energy*, *time* and *mass*.[147]

[147] See Ch. 2.

Every object or action can be described with these elements. Modern physics explains their integration to form the world as being controlled by Natural Law, but the ancient philosophers argued that there had to be a source or giver of Law that preceded the physical manifesting.

The ability of the building blocks to respond to a vision requires some form of intelligence and energy within each building block. The outcome is that the final change results in the object or action shining forth the characteristic that was first only envisioned. The *Rudrāyamala* continues to describe the power of the vision which can change the basic elements to change the world.

8. The building blocks or elements are without beginning but integrate together to manifest the mental deliberations and vision called a *mantra* of a ruling god.

Literal: Without beginning, they proceed to become named and perceived creations, truly the deliberations with knowledge and beauty of a supreme being.

amūlā tatkramāj jneyā kṣāntā sṛṣṭr udāhṛitā /
sarveṣām eva mantrāṇām vidyānāṃ ca yaśasvini // 8

amūlā: without root. ***tat***: then. ***kramāj***:proceeding. ***jneyā***: to be perceived. ***kṣāntā***: enduring. ***sṛṣṭa***: created. ***udāhṛtā***: called, name. /	***sarveṣām***: supreme being. ***eva***: truly. ***man***: to think one's self. ***tra***: protector. ***mantrāṇām***: to deliberate, envision. ***vidyānāṃ***: knowledge. ***ca***: and. ***yaśasvini***: illustrious. // **8**

Comment: The special power of the higher mind, *mantra*, can be compared to the older psychological term "conation" which is the mental creation of a future before it becomes manifested. That which is yearned for before it appears is *mantra. Mantra* is far more than a wish; it is the mental creation of what is to be.

Recent efforts at Princeton University have served to prove that individuals can influence small changes in precision devices such as clocks by "willing" a change. Since the force of *mantra* is weak it must be prolonged and continually directed. As such it requires the continuous step-by-step efforts towards its fulfillment which many individuals are thoroughly familiar with. In the process the elements are used to make the desired world manifest. As the personal world slowly changes and becomes constantly more perfect and filled with beauty it is best described as becoming magnificent.

Every evolutionary world that is chosen or created within the present moment has no earlier origin or beginning. Anything that has the sense of reality is defined by *mantra* and becomes manifested as shining knowledge or divine experience.

9. The *yoni* is credited with being the source of the manifesting of the dedication of the heart.

Literal: This *yoni* is defined as bestowing and protective. The hidden *yoni* bestows union, prosperity, joy and manifests the inner yearnings.

iyaṃ yoniḥ samākhyātā sarvatantreṣu sarvadā /
caturdaśayutaṃ bhadre tithīśānta samanvitam // 9

iyaṃ: this.
yoniḥ: source of generation.
samākhyā: calculated, enumerated.
sarva: entire.
tan: extent.
tra: protection.
sarvadā: all bestowing. /

cat: hidden.
urda: cheerful, upper.
śa: bestowing.
yuta: united.
bhadre: prosperity.
tithī: ***titha***: love, fire.
śānta: joy.
samana: union.
vitam: manifested desire. // **9**

Comment: The masculine noun *yoni* refers to the center of the perineum which when activated swells and stimulates the *hridaya* or (lower) heart. (There is the similar feminine noun *yonī* in Indian literature which is the female pudendum and is also a creative and responsive center.) The stimulation of the *yoni* produces a noticeable swelling called the *kanda* or *lingam* in men and women. The swollen *lingam* was depicted in icons as a source of masculine creative power in Greece as well as India.

10. The androgynous nature of the Creator *Brahma* as well as the inner *Rudra* (or *Brahmā*) exists in the heart, located between the thighs. If this androgynous nature cannot be found, an individual cannot find enlightenment.

Literal: The third nature of *Brahma* is activated by the Soul pressing the heart between the thighs. Those who do not have the existence as a *yoginī* (or the state of androgyny), as did the god *Rudra*, cannot break forth.

tṛtīyaṃ brahma suśroṇī hṛdayaṃ bhairavātmanaḥ /
etannāyoginījāto nārudro labhate sphuṭam // 10

tṛtīyaṃ: the third.
brahma: a god.
su: creating, activating.

etan: this. ***nā***: no, not so.
yoginī: female yogi.
jāto: produced.

śroṇī: loins. ***hṛdayaṃ***: heart. ***bhairavā***: *Rudra*. ***tmanaḥ***: Soul. /	***nārudro***: ***nā***: no, not so. ***rudra***: god half male and half female. ***labhate***: obtaining. ***sphuṭam***: appear suddenly. //**10**

Comment: The Creator Gods must contain both the vision of what is to be manifested as well as the source of manifesting power. The third nature is the combining of both of the powers in independent creative centers such as within enlightened individuals. An enlightened individual therefore contains both an inner masculine as well as a feminine power to manifest the vision. An individual who has both active masculine and feminine powers is called a *yoginī* and described as having special mystical powers. One myth of the androgynous god *Rudra* was highlighted by how, after being locked deep in a well, he escaped through the creation of his own freedom, which is of course highly suggestive of escaping our own entrapment with life.

11. **The lower heart is the dwelling place of *Rudra* or the power of Love and provides the energy for maintaining union with as well as liberation from the outer world. To create a new world, a vision or *mantra* must be created, followed with *mudra* which transforms a vision into reality.**

Literal: The *hridaya* is the dwelling place of *Rudra*, the god of gods, and is the source of union with liberation at the same time. Ascending (beyond) is accomplished with the uniting of the great *mantra* and *mudra*.

hṛdayaṃ devadevasya sadyo yogavimuktidum /
asyoccāre kṛte samyaṇ
mantramudrāgaṇo mahān // 11

hṛdayaṃ: heart, center of feelings. ***devadevasya***: god of gods. ***sadyo***: ***sādhya***: *Rudra*, god of Love. ***yoga***: union. ***vimukti***: set free. /	***asyo***: become. ***uccāre***: to rise. ***kṛte***: making, doing. ***samyaṇ***: right, correct. ***mantra***: vision: ***mudra***: manifested vision. ***agano***: attain. ***mahān***: great. // **11**

Comment: The term "god of gods" for *Deva* or *Rudra* is easier to understand from the early Greek definitions of the inner masculine god *Eros* who also had a form of *Zeus* in the heavens above. The "god of gods" rules the powers or lesser gods within the body. *Deva* is the manifesting god of the yearnings of *Devī* or the Soul.

Stating that *Deva* is the source of union with liberation might seem contrary to modern concepts which views liberation as being free from uniting forces. The liberation, however, refers to the ability of the individual to choose a completely different world and self while the union refers to the ability to become interactive or an integral part of that new world.

12. Becoming an envisioned new person requires the full acceptance of the new manifested image or *mudra*, not as new, but as an old, highly loved image that was created or chosen in the past as a *mantra*.

Literal: A perfected and expanding *mudra* or manifested vision of the self is obtained by bestowing and facing the arrival within one's body of the chosen characteristics obtained with a strong effort as well as in the moment of reception creating a loving tender recollection of the *mudra*.

**sadya sanmukhatam eti svadehāveśalakṣanam /
muhūrtam smarate yas
tu cumbake nabhimudritaḥ // 12**

sadya: to be perfected. ***san***: bestow. ***mukhatam***: confronting, facing. ***eti***: arrival. ***svadehā***: one's body. ***veśa***: to assume an appearance. ***lakṣana***: relating to or acquainted with characteristic signs or marks. /	***muhūrtam***: a moment. ***smarate***: loving recollection. ***yas***: exert. ***tu***: to make strong, efficient. ***cumb***: to touch closely or softly. ***nabhi***: expanding. ***mudritaḥ***: obtained *mudra*. // **12**

Comment: This stage is difficult for modern individuals, since most people consider that they have only one self, namely the identity that became solidified as a child and reinforced by friends, family and associates. It is also difficult to accept the concept that a different self can be accepted in the same nature as putting on another pair of clothes. However, almost all people do in fact change their roles or selves as they "act" like a parent, a family member, a supervisor or a friend.

13. At that time the present image or role of the self is lost and replaced with an image or *mantra* created in the past for the future. It rises from the depths in the present under a strong drive or need.

Literal: At that time when one procures annihilation (of the old *mudra*), the body can be molded with the spirit of a new *mantra-mudra*. Which was created in an envisioned future but is driven to rise from the past to become manifest in the present.

sa badhnāti tadā dehaṃ
mantramudrāgaṇaṃ naraḥ /
atītanāgatānarthān pṛṣṭo'sau kathayatyapi // 13

sa: procuring. ***badh***: ***vadha***: annihilation, disappearance. ***tadā***: at that time. ***dehaṃ***: to mold the body. ***mantramudrā***: *mantra-mudra*. ***āgaṇaṃ***: arriving. ***nara***: masculine, male, eternal spirit. /	***atītānā***: ***atita***: gone by. ***anāgata***: not come. ***narthān***: ***nṛt***: to act. ***pṛṣṭo***: rising from behind. ***asau***: underneath. ***katha***: talk, description, image. ***yatyapi***: ***yatya***: driven. ***api***: very. // **13**

Comment: The best example of this is how people will relate how they suddenly forgot their assumed limitations and became heroic or managed to do something that was completely unexpected. Invariably, they will state how they first saw themselves doing the impossible or unexpected and then that they had no choice but to manifest what they envisioned.

The perception of the future act was not of a conscious effort but arose from the depths of their consciousness or from a *mantra*. Similarly, the forgetting of who or what they were and the concept of their limitations is not of conscious effort.

14. Using *mantra* and *mudra* one can reach for and manifest the spirit of any desired god or divine nature.

Literal: The androgynous power of *Rudra* can be thrust forward to manifest the spirit of any god which becomes overpowering and enveloping.

**praharād yad abhipretaṃ devātarūpam uccaran /
sākṣāt paśyatyasandigdham
ākṛṣtam rudraśaktibhiḥ // 14**

prahara: to thrust or move forward.	***sākṣāt***: overpowering, with one's own eyes.
yad: which.	***paśya***: beholding.
abhi: approach.	***digdham***: covered, anointed.
pretaṃ: deceased.	***ākṛṣtam***: attracted.
devatā: divinity.	***rudraśakti***: by the power of *Rudra*. // **14**
rūpam: form.	
uccaran: ascend, rise, issue forth, go forth. /	

Comment: The Greeks were convinced that a hero managed to assimilate the inner nature of a god, but no records survive as to how this was done other than the chosen god had to be fed ambrosia (or the proper hormones had to be generated). The Greeks did not use *mantra-mudra* but did describe the same relationship as "sun-moon." The sun became the source of vision or *mantra*, and the moon became the manifesting or *mudra* of the sun's power.

Mantra-mudra has become lost in the modern world's denial of the metaphysical *mantra* and its emphasis upon the physical *mudra* or the outer characteristics of an individual. If there is no underlying *mantra* to an action it can only be an "act" or a conditioned response without any assumed reality or reason for the role. Therefore, in modern society individuals are judged by their obedience, training or titles rather than their inner abilities.

The modern concept of changing is to manipulate effects rather than causes.

15. Ultimate truth is found only with Love[148] of such power that the bowels churn for its attainment.

Literal: The ultimate truth of heaven is brought forth by churning produced by striving after Love by means of protection and of the carrying across and joining with all of the great strength of *Rudra* or inner Love.

praharādvayamātreṇa vyomastho
jāyate smaran /
trayeṇa mātaraḥ sarvā
yogīśvaryo mahābalāḥ // 15

prahara: to thrust forward. ***advaya***: ultimate truth. ***mātreṇa***: only by. ***mathra***: churning. ***eṇa***: by means of. ***vyomastho***: being in the sky, heavenly. ***jāyate***: ***ja***: produced by. ***yate***: strive after. ***smara***: love, the god of love, *Rudra*. /	***tray***: ***trai***: protect. ***eṇa***: by means of. ***mā***: traverse. ***taraḥ***: carrying across. ***sarvā***: all. ***yogi***: join, unite. ***īśvarya***: the god of love, *Rudra*. ***mahābalāḥ***: great strength. // **15**

Comment: The god of Love can be compared with the Greek inner god *Eros* who was the predecessor of Cupid. *Eros* contained the fire of love that led individuals into heroic (from *eros*) actions. The love of *Eros* or Love was credited with the ability of people to change and become subject to or manifest their Love such as a person loving knowledge could become a "philosopher"[149] or lover of knowledge. *Eros*, like *Rudra* and Love, was known to require an activating or immortalizing nectar called the drink of the gods, *ambrosia*,

[148] See Ch. 7.

[149] Greek: *philo*, "love" + *sophia*, "knowledge"

which is the same as the *soma* or *amrita* to be introduced shortly.

16. A retinue of inner supporting powers arrives, brought together by the power of Love.

Literal: A hero and the perfected powerful chief of heroes, a portion of the total arrive and come together as givers driven on by means of *Rudra*.

vīra vīreśvaraḥ siddhā
balavān chākinīgaṇah /
āgatya samayaṃ datvā
bhairaveṇa pracoditāḥ // 16

vīra: hero. ***vīreśvaraḥ***: chief of heroes. ***siddhā***: perfected. ***bala***: possessing power. ***cha***: a fragment. ***akin***: ***ākīm***: from. ***gaṇa***: a troop deity (esp. pl. the followers of *Śiva*.) /	***āgatya***: having arrived or come. ***samayaṃ***: to come together. ***datvā***: gift, a giver. ***bhairaveṇa***: by means of *Rudra*. ***pracoditāḥ***: driven on. // **16**

Comment: (No comment needed.)

17. One must take great effort to first perceive and then obtain total perfection. To be without sin, one must find wisdom and perfection in the past, present and future.

Literal: Who with perceived great effort fulfills and gains the highest perfection. Being without sin, perfected in the past and future, is perfected in the present and wise.

yacchanti paramām siddhiṃ
phalaṃ yadvā samīhitam /
anena siddhāḥ setsyanti
sādhayanti ca mantriṇaḥ // 17

yac: ***yad***: who.	***anena***: sinless.
chanti: ***chad***: appear, pleased, cover.	***siddhāḥ***: perfected in the past.
paramām: highest, primary.	***setsyanti***: perfection in the future.
siddhiṃ: perfection, fulfill.	***sādhayanti***: perfected in the present.
phalaṃ: benefit, result.	***ca***: and.
yadvā: perception.	***mantriṇaḥ***: wise. // **17**
samīhitam: great effort to obtain anything. /	

Comment: Wisdom demonstrates that every step in evolving leads to a defined goal, even though at the moment the step may appear to be detrimental to evolution. It is common to hear perfected individuals state how their perceived setbacks always turned out to be positive experiences. If individuals are content with their present state of life, then every past experience must be accepted as perfect since it led to the present.

This verse teaches that an individual can sin if any anticipated or planned activity is not perceived as perfect. This requires its comparison with the past, present and future. This verse is supported by the Greek word for sin, *hamartano*, which has the meaning of "missing the mark" or "missing a personally chosen goal". A further concept of sin can be obtained from the Greek word *dikaios* which was applied to an individual without sin. This word meant "equal, even or well-balanced," as well as "legally exact and precise." Such an individual was called righteous, able to follow the right path leading to personal perfection.

18. Obtaining a perfect life requires using the power of *Rudra* or Love to define the goal and then churning of the guts or flesh. The churning of the flesh combined with the merger of the *mantra* of becoming a hero will fulfill any state of mind or body to obtain any desire. However, this cannot be done by the conditioned self-control or restraints.

Literal: Success in the desire for anything is in the nature of *Bhairava* or *Rudra* with the universal churning of the flesh, with the entrance of the hero *mantra*, any state of mind or body, not by controlling restraints.

yatkincid bhairave tantre sarvamasmāt prasiddhyati /
mantravīryasamāveś aprabhāvān na niyantriṇā // 18

yatkincid: desire for anything whatever.
bhairave: *Rudra*.
tantre: rule, essence.
sarva: universal, for everyone, all.
mas: flesh. ***mat***: churn.
pra: great. ***siddhya***: success. /
mantra: ***vīrya***: hero.
samāveś: entering together.
apra: to fulfil.
bhāva: any state of mind or body.
na: not.
niyantriṇā: controlling restraining. // **18**

Comment: (No comment needed.)

19. There is an unexpected addition of swelling that is felt in the perineum with churning and mental concentration. This protuberance speeds the division between the old and the new heroic true state. The protuberance also gives partial fulfillment to the initiation and continual attainment of enlightenment.

Literal: The annexing of an unexpected fleshy expelled excrescence speeds the division to the going to the strong and true state of existence. This initiates and gives partial fulfillment of becoming a permanent *yogī*.

adṛṣṭamaṇḍalo'pi evaṃ yaḥ kaścid vetti tattvataḥ /
sa siddhibhāgbhaven nityaṃ
sa yogī sa ca dīkṣitaḥ // 19

adṛṣṭa: unexpected, unseen. ***maṇ***: ***maṇi***: fleshy excrescence. ***dala***: expel. ***api***: annexing, to join to, pour out.	**sa**: giving.
evaṃ: thus.	***siddhi***: accomplishment, fulfillment. ***bhāg***: a part, portion, share. ***bhaven***: becoming, being.
yaḥ: ***yah***: to speed, be quick.	***nityaṃ***: continual, perpetual.
kaś: to go. ***cid***: to divide.	***sa***: giving.
vetti: ***vid***: to be strong.	***yogī***: ***yogini***: female *yogi*.
tattvataḥ: true or real state. ***āta***: going. /	***ca***: and.
	dīkṣitaḥ: initiated into. // **19**

Comment: The described swelling or protuberance can be felt to some extent with deep crying, laughing, coughing or strong exhalations. There are a number of Sanskrit words describing this protuberance or swelling with the most common being *lingam* (sign of attainment or enlightenment) as well as *kanda*. The initiation or *dīksha* corresponds to the attainment of a higher power.

20. A mighty strength in knowledge is acquired by the uniting of knowledge gained by churning and that gained by practical experience and learning. That union provides total knowledge without error.

Literal: Without error bringing into existence the knowledge acquired by churning and that gained by practical experience. That mighty assembly (of knowledge) is stimulated to come into a united existence.

anena jñatamātreṇa jñāyate sarvaśaktibhiḥ /
śākinīkulasāmānyo bhaved yogaṃ vināpi hi // 20

anena: without error, sinless. ***jñāta***: known. ***mātreṇa***: by churning. ***jñā***: to know. ***yate***: to join. ***sarva***: all. ***śakti***: adherence, practical, ability. ***bhiḥ***: by means of. /	***śākinī***: mighty, female. ***kula***: assemblage. ***sāmānyo***: in common, equal. ***bhaved***: ***bhu***: coming into existence. ***yogaṃ***: union. ***vina***: discipline. ***api***: uniting to. ***hi***: to stimulate or incite to. // **20**

Comment: This verse is very important since it states that Truth can only be known if the metaphysical forces and causes are considered along with the facts gained from the manifested world. In a practical sense, this is stated as being certain about some conclusion since your inner gut feelings and your conscious thoughts both agree.

21. Without worship and study one knows all that is required and no longer requires religious worship or sacrifice. A consuming fire breaks the bondage of illusion producing the perception of the highest God.

Literal: No worship or study, procures all that is knowable and is victorius over sacrificing. The highest God is perceived with the fiery consumption of illusion.

avidhijño vidhānajño jāyate yajanaṃ prati /
kālāgnim āditaḥ kṛtvā māyāntaṃ
brahmadehagam //21

avidhi: no workship. ***jño***: study, knowing. ***vidhā***: procure, bestow. ***anājñā***: surpassing all that has ever been known. ***jāyate***:victorious. ***yajanaṃ***: sacrificing. ***prati***: towards. /	***kālāgnim***: consuming fire. ***ādita***: not bound, beginning. ***kṛtvā***: producing. ***māya***: illusory or supernatural power. ***anta***: issue, event. ***brahma***: the highest god. ***deha***: body, person. ***gam***: perceive. // **21**

Comment: Churning consumes the imagined reality of conditioned thought, producing a perception and access to the highest level of knowledge.

22. There are three evolutionary powers that must be maintained and utilized throughout one's whole life. These are to excel in everything that is done, to have a vision as to direction, and to exert one's self in the manifesting of the vision. In order to follow this path, one must be prepared to do anything.

Literal: Understand that the auspicious three personal powers (preeminence, *mantra*, and effort) include everything through the beginning, end of life and future life. Therefore, the inward subsistence of doing anything becomes a pure path for the one who perseveres.

śivo viśvādyanantāntaḥ paraṃ śaktitrayaṃ matam /
tadantarvarti yatkincit
śuddhamārge vyavasthitam // 22

śivo: auspicious. ***viśva***: all, entire, whole.	***tad***: then. ***antar***: inward. ***varti***: subsistence.

***dyanant*: *adyanta*:** beginning and end.
***anta*:** end (in time), end of life, death.
***para śakti traya*:** performing the three energies (*prabhutva*: personal pre-eminence. *mantra*. *utsāha*: effort.) with all one's might.
***mata*:** understand. /

***yat kincit*:** doing anything.
***śuddham*:** purity.
***mārge*:** proper course.
***vyavasthiti*:** perseverance in. // **22**

Comment: This verse is contrary to modern social dogma, since one's own goal becomes paramount as well as one's own inner powers to attain it. It should be remembered, however, as sages have reported throughout the millennia, that when individuals satisfy their visions (not desires), society is improved and evolves.

23. In just a short moment, supreme knowledge can be obtained from the inner *Rudra* or god of Love.

Literal: In one moment of enlightenment supreme, knowledge is consumed. That highest auspicious all-knowing *Rudra*.

aṇurviśuddham acirāt aiśvaraṃ
jñanam aśnute /
taccodakaḥ śivojneyah
sarvajnaḥ parameśvaraḥ // 23

***aṇu*:** moment. ***viśuddham*:** enlightened.
***acirāt*:** not long.
***aiśvara*:** supremacy.
***jñanam*:** knowledge.
***aśnute*: *aśnat*:** consuming. /

***taccodakaḥ*: *tac*: *tad*:** that. ***codakaḥ*:** impelling, excitation, inner power.
***śivojneyah*:** knowledge of what is fortunate, auspicious to be known.

	sarvajnaḥ: all knowledge. ***parama***: highest. ***iśvaraḥ***: inner god, *Rudra*. // **23**

Comment: This is a common experience of following the inner deep feelings of Love. The ancient world used the expression, "hit by lightning" which might be more common today as "hit over the head" when suddenly a whole and detailed view of the object of Love is understood.

24. The power of Love in the heart is like a seed which, when nourished, can grow into a great tree.

Literal: The all-pervading *Rudra*, fully satisfied, radiant and abiding in his own essential nature is like the great banyan tree contained within the energy of its seed.

sarvago nirmalaḥ svacchas tṛptaḥ
svāyatanah śuchiḥ /
yathā nyagrodhabījasthaḥ
śaktirūpo mahādrumaḥ // 24

sarvago: all pervading Soul. ***nirmalaḥ***: pure, shining. ***svacchas***: clear. ***tṛptaḥ***: satiated. ***svāyatanah***: one's home, resting place. ***śuchiḥ***: purity. /	***yathā***: in like manner. ***nyagrodha***: the great banyan tree. ***bīja***: seed. ***sthaḥ***: to be founded upon. ***śaktirūpo***: energy of seed. ***mahādrumaḥ***: a great tree. // **24**

Comment: (No comment needed.)

25. In like manner, the inner seed in the heart contains the elements that produce the total world which contains the original vision. Truly perfection or liberation from ignorance is obtained with the inner rising knowledge and truth.

Literal: In this manner, the seed existing in the heart contains the aggregate of all creations, both motionless and moving. Truly, liberation is attained in the casting up of knowledge and truth.

tathā hṛdayabījasthaṃ jagadetaccarācaram /
evaṃ yo vetti tattvena tasya nirvānagāminī // 25

tathā: in that manner. ***hṛdaya***: heart. ***bīja***: seed. ***sthaṃ***: existing. ***jagat***: world. ***eta***: approach. ***caracara***: the aggregate of all creation. /	***evaṃ***: truly. ***yo***: ***yuj***: united. ***vetti***: knower. ***tattvena***: truth. ***tasya***: cast up. ***nirvāna***: beatitude, liberation, oblivion. ***gāminī***: attaining, going. // **25**

Comment: The manifesting of an envisioned change in either the self or world must contain a change in everything which is even remotely connected to the vision. For instance, in a dream everything fits and functions together in complete union. A desired change, in contrast, is limited to the image of the change but not the surrounding world.

26. Self-justification, anointings and religious ceremonies are abandoned. Likewise, everything begins within the embodied concealed heart.

Literal: One becomes free from consecrations, religious anointings and meditations. In like manner, the initial first part to bring a new state begins with the material form of the concealed hidden center.

dīkṣa bhavatyasaṃdigdha tilājyāhutivarjitā /
mūrdhni vaktre ca hṛdaye guhye
mūrtau tathaiva ca // 26

dīkṣa: consecrate one's self. ***bhava***: coming into existence, origin. ***saṃdigdha***: anointings. ***tila***: small particle. ***ājya***: religious markings, oblations. ***ahuti***: meditations, invocations. ***varjit***: free from. /	***mūrdhni***: ***mūrdhan***: the head, the first part of anything. ***ni***: to bring into any state or condition. ***vaktre***: ***vaktra***: mouth, beginning, the initial or first term of a progression. ***ca***: as well as, and. ***hṛdaye***: center or heart. ***guhye***: concealed, sexual organs. ***mūrtau***: material form. ***tathaiva ca***: in like manner. // **26**

Comment: A change occurs from being concerned about what others think or what is proper to becoming the individual who fulfills the vision of what one has a love to become.

27. The nipples and breasts are stimulated to find the power that can manifest the inner vision.

Literal: Having deliberated to bestow goodness, the nipple and breast are clenched and caressed one-by-one to have power and direction to reach and make the condition of the vision.

nyāsaṃ kṛtva śikhām baddhvā
saptaviṃśatimantritām /
ekaikaṃ tu diśām bandhaṃ
daśānām api kārayet // 27

nyāsaṃ: placing, impressing, applying.
kṛtvā: having done.
śikhām: nipple, spike, projection, end or point.
baddhvā: having clenched, joined, tied.
saptaviṃśati: ***sap***: caress, seek. ***ta***: the breast, a tail. ***viṃ***: ***vin***: bestow. ***śati***: goodness: ***mantritām***: deliberated./

ekaikaṃ: one-by-one.
tu: to have power.
diśām: direction.
bandhaṃ: ***bandh***: to unite.
daśān: ***daśā***: condition, wick, ten. ***nām***: vision, name, appearance.
api: reaching to.
kārayet: ***kṛ***: a doer, maker. // **27**

Comment: This verse may well be the secret stage of *Yoga* that the masters would teach their awakening students by allowing them to observe themselves in meditation. It certainly would be a far better method of teaching than attempting to describe it. This verse is likewise an example of how the inner meaning of some verses was hidden. The words *sap ta viṃ* and *śati* can be compounded together to also produce the word for "twenty-seven" which is much easier to read than to separate the words.

The word *śikhām* has the fundamental meaning of "something pointed," which leads to its usage to describe the nipple. It also is used to describe such things as a peacock's crest or comb which probably suggested to some translators that it could be applied to the tying of a lock of hair for a religious symbol.

This leads to the common translation that a tuft of hair is to be tied with twenty-seven *mantras*, which may be representative of the problems encountered with the *Rudrayāmala*. However, it is only when the Sanskrit is unscrambled properly that a sensible meaning appears using the literal meanings of the words with no externally added words.

28. The stimulation produces a swelling in the middle of the perineum which in turn produces a vibration in the whole body removing the effects of any passions as obstacles. The nipples can also be considered as producing a mystical fluid which permeates the whole body.

Literal: The protective control produces swelling procuring the state of vibration removing obstacles and passions. The nipples are considered as producing a divine fluid permeating throughout all the body.

tālatrayam purā dattvā saśabdaṃ vighnaśāntaye /
śikhāsaṃkhyābhijaptena
toyenābhyukṣayet tataḥ // 28

tāla: measure, control. ***trayam***: ***trai***: to protect. ***purā***: swelling, stronghold. ***dat***: ***da***: producing. ***tvā***: the state of being. ***sa***: procuring. ***śabda***: vibrations. ***vighna***: removal of obstacles. ***śāntaye***: reducing passions. /	***śikhā***: nipples. ***saṃkhyā***: to be considered as. ***abhija***: produced all around. ***ap***: divine fluid. ***tena***: in that manner. ***toya***: fluid. ***ena***: it, this, that. ***abhyukṣ***: sprinkling over, wetting. ***aye***: in this manner. ***tataḥ***: spread, diffused. // **28**

Comment: (No comment needed.)

29. There is an abundant discharge of fluid in both sexes from flat-chested to the full-breasted. The production begins with manifesting the imagined nourishing fluid.

Literal: An abundant discharge a step for all genders of either bare ground or mountain, sitting and imagining the nourishing moisture working in that direction as a seed.

puṣpādikaṃ kramāt sarvaṃ
liṅge vā sthaṇḍile thavā /
caturdaśābhijaptena puṣpeṇāsanakalpanā // 29

***puṣpā*: *puṣpāka*:** discharge.
***adika*:** abundant.
***krama*:** a step, going, course.
***sarvaṃ*:** all, whole.
***liṅge*: *liṅga*:** gender, the sign of gender.
***vā*:** either.
***sthaṇḍile*:** bare ground.
***thavā*: *tha*:** a mountain.
***vā*:** or. /

***caturda*:** four. ***cat*:** to go.
***ūrd*:** to play.
***aśābhij*:** without a seed.
***śā bhijā*:** with a principle or source.
***aptena*: *ap*:** work.
***tena*:** in that direction.
***puṣpeṇā*: *puṣ*:** nourishing.
***peṇā*: *pheṇā*:** foam, moisture, embrace.
***āsana*:** sitting.
***kalpan*:** fashioning or performing, forming in the imagination, inventing. // **29**

Comment: This verse counters the later universal consideration of females as not possessing an inner power or perhaps not even a soul. The size of the nipple and breast is also stated to be unimportant. The imagining of an inner flow of a fluid would be described today as positive feedback in learning to secrete it. Many people describe the effect as like a drug rush which is known to be enhanced with concentration upon the effects.

30. Under those circumstances and with a strong sacrificing of the self (to the feelings), the emission of *soma* is renewed with a union of the perineum or buttocks with the seat and with protracted sitting.

Literal: Under those circumstances with protracted sitting, *soma* is renewed. Strong emission is attained with a union of the buttocks from behind and self-sacrifice.

tatra sṛṣṭiṃ yajed vīraḥ punarevāsanaṃ tataḥ /
sṛṣṭiṃ tu sampuṭīkṛtya paścād yajanam ārabhet // 30

tatra: under those circumstances.	***sṛṣṭiṃ***: emission.
sṛṣṭiṃ: emission, creation.	***tu***: to be strong.
yajed: ***yaj***: to consecrate.	***sampuṭī***: (a kind of sexual union). ***sam***: union of.
vīraḥ: hero or an intoxicating beverage (*soma*).	***pūta***: the buttocks.
punar: to restore, renew.	***kṛtya***: action.
eva: as long as.	***paścād***: ***paścāt***: behind.
asanaṃ: sitting.	***yajanam***: the act of performing a sacrifice.
tataḥ: protracted. /	***ārabhet***: ***ārabh***: reach, attain. // **30**

Comment: (No comment needed.)

31. In order to manifest a created *mantra*, it is necessary to allow the body to become filled with a creative comforting feminine force. The feminine nature is then actuated by stimulating or rubbing the breast.

Literal: Seeking to obtain complete reality, submit to a great feminine power by caressing and inciting the breast to bestow and gain the remembered *mantra*.

sarvatattva susampūrṇaṃ sarvābharaṇa bhūṣitām /
yajed devīṃ maheśānīṃ
saptaviṃśati-mantritām // 31

sarva: whole. ***tattva***: truth or true reality.
su: to impel. ***sampūrṇaṃ***: filled with.
sarvā: whole. ***bharaṇa***: maintaining.
bhūṣitām: ***bhūṣ***: to seek. ***itām***: ***ita***: obtain. /

yajed: ***yaj***: consecrate.
devīṃ: a female deity.
maheśānīṃ: ***maheśāna***: great lady.
saptaviṃśa: ***sap***: caress. ***ta***: the breast. ***viṃ***: ***vin***: to bestow. ***śati***: one hundred, gaining, obtaining. ***antritām***: ***mantra***: ***ita***: returned remembered. // **31**

Comment: This verse could also be expressed as the common expression that an individual gives up attempting to know or change a situation and rather sinks into a state of being enveloped with a feeling of security, warmth and love. In seeking this state, it is common to pull the knees up to the chest, and clench or tightly rub the chest. This state of mind and body is also found in uncontrolled sobbing.

32. The manifesting of the virtuous *mantra* requires the submission to a cover of gods or forces which are able to overwhelm the Soul.

Literal: That virtuous expansion, according to ability, is accompanied by the cover of gods who when honored overwhelm the vital body and manifest (the *mantra*).

tataḥ sugandhipuṣpaistuyathāśaktyā samarcayet /
pūjayet parayā bhaktyā ātmānaṃ ca nivedayet // 32

tataḥ: that ***sugandhi***: ***sugandha***: virtuous, fragrant. ***puṣpaistu***: ***puṣpa***: expanding. ***yathāśaktyā***: ***yathāśakti***: according to power or ability. ***samarcayet***: ***samar***: accompanied by the gods. ***cayet***: ***caya***: a cover. /	***pūjayet***: ***pūjayāna***: honoring. ***parayā***: ***parayatta***: overwhelmed by. ***bhaktyā***: ***bhakta***: engaged in. ***ātmānaṃ***: the Soul. ***ca***: and. ***nivedayet***: manifesting, reporting or relating. // **32**

Comment: This describes the mental state which goes with the above state of the physical body as the mind not only gives up control of the body, but also opens to inner healing or guiding forces. This is the state normally assumed, perhaps unconsciously, by those who are readying themselves to "sleep on" a problem and are expecting an answer or insight.

33. The reached for, dedicated to, or prayed for result is finally obtained by the activation of the inner power of Love which integrates the past and future with the reality of the present.

Literal: Truth and perfection is found by both feeding and then yielding to the inner forces. The dedication to truth initiates the power of Love which then is the source of bringing forth great actions.

**evaṃ yajanam ākhyātam agnikārye pyayaṃ vidhiḥ /
kṛtapūja vidhi samyak samaran
bījam prasiddhyati // 33**

evaṃ: thus
yajanam: the act of sacrificing.
ākhyātam: made known, manifested.
agnikārye: feeding the inner sacrificial fire.
pyayaṃ: approach.
vidhiḥ: law, correct manner, truth. /

kṛta: made. ***pūja***: to initiate, consecrate.
vidhi: prescribed act or rite.
samyak: right, correct, perfection.
smaran: act of love or union.
bījam: beginning, seed.
prasid: accomplished.
pra: great. ***sid***: to turn out well, have success.
yati: disposer. // **33**

Comment: (No comment needed.)

34. Breaking free from the bondage of time, the fluid extends to the love and fire in the heart. A portion of the fluid supports the inner reigning *Aphrodite* or Soul.

Literal: Free from beginning or ending, the fire of the seed expands and goes to the center of the body, portion of the nectar bringing within the lotus, or inner heart, the eternal reigning *Lakṣmī* or Soul.

**ādyantarahitaṃ bījam vikasat tithimadhyagaṃ /
hṛtpadmāntargataṃ dhyāyet
somāṃśaṃ nityam abhyaset // 34**

adyanta: beginning and end. ***rahita***: separated or free from.
bījam: source, seed, the position of the arms of a child at birth.
vikasa: extend, increase,

hṛt: bringing.
padma: lotus, *Lakṣmī*.
antar: within.
gata: going.
dhyāyat: thinking, meditation, imagining.
soma: fluid. ***amśa***: portion.

to open, expand. ***tithi***: ***titha***: fire, love. ***madhya***: central, middle of body. ***ga***: going. /	***nityam***: eternal. ***abhyas***: reign over. // **34**

Comment: The inner goddess *Lakṣmī* is the same as the Greek goddess *Aphrodite*. Both goddesses are described as beautiful as well as the source of success and attainment. The inner generated fluid of *ambrosia* (Greek) and *amrita* or *soma* (Sanskrit) are both described as food for the inner gods. The Soul is not the same as the god of Love, *Eros* (Greek) or *Rudra* (Sanskrit).

35. Going, going with Love. Love without a defect, to strive after a place of strength, to arrive at the goal. Becoming an immovable living supernatural being, approaching knowledge of all and with no doubt about it.

Literal: (No literal needed.)

yānyān kāmayate
kāmā-stāṃstācchīdhādhramavāpnuyāt /
asmāt pratyakṣatām
eti sarvajñatvaṃ na saṃśayaḥ // 35

yānyān: going going. ***kāma***: love. ***yat***: strive after. ***stāṃ***: ***sthaman***: place seat, strength. ***sta***: to stand firm. ***acchidhra***: without a hole. ***mava***: ***nava***: new. ***ap***: water, work, to arrive at one's aim or end, become filled. /	***asmāt***: stone. ***prat***: ***prati***: to go towards. ***yakṣa***: a living supernatural being. ***tām***: to become immovable. ***eti***: approach. ***sarva***: everything. ***jña***: to know. ***tvaṃ***: thee. ***na***: ***saṃśayaḥ***: no doubt about it. // **35**

Comment: (No comment needed.)

36. This manifested *mantra* bursts forth from the combined masculine and feminine powers to attain all the knowledge and powers.

Literal: (No literal needed.)

evaṃ mantra phalavāptirity etad rudrayāmalam /
etad abhyāsatah siddhiḥ
sarvajñatvam avāpyate // 36

evaṃ: this. ***mantra***: thought. ***phala***: to bring to maturity. ***avāpti***: ***avapti***: attainment, obtaining. ***vā***: like. ***āpti***: connection, sexual intercourse. ***etad***: this. ***rudrayāmalam***: combined masculine and feminine powers. /	***etad***: this. ***abhya***: towards. ***asa***: the lower part of the body, behind, posterior. ***siddhiḥ***: attainment. ***sarvajña***: knowing all. ***tvam***: thee. ***avāpya***: ***avapya***: to be obtained. // **36**

Comment: This verse states that through Love, anything can be gained.

Appendix
Measuring Improved Response Time with Girding or *Bandha*

A simple experiment was performed with a group of people who had been practicing some of the ancient *bandha* exercises (See Chapter Eight) in terms of their reaction time. This experiment was devised to prove that those who practice the ancient exercises of *bandha* and churning have a much quicker response time and explains why the martial arts came forth from the early schools of *Yoga*.

A measurement of the time it took a group of 15 middle-aged people using the prescribed exercises to respond to the sound of a clap was compared with a control group of about the same average age and a control group of children. The test required keeping the eyes closed to avoid any visual clues and keeping their hands at the side to equalize the required distance to move the hands and then to clap the hands as soon as possible following the sound of a clap. A sound engineer analyzed a sound recording to evaluate the actual time of response.

The two control groups consisted of a group of 18 children, 12 to 13 years old and a group of 14 middle-aged men and women. Two times were measured, one for the fastest individual response and the second for the time between the first and the last clap or the time interval of the groups response which gave an indication of the uniformity of the group's response.

In seconds	Adults	Girded Adults	Children: 12-13 Years
Fastest Time	0.320	0.170	0.240
Time Interval	0.250	0.110	0.140
Average Time	0.445	0.225	0.310
Slowest Time	0.570	0.280	0.380

This study is quite exciting since it indicates that girding practices counter the normal drop off in response time with aging, with for instance the conclusion that a seventy-year-old girding student had a response time faster than an average 12 year old. It also shows that the average response time takes twice as long for people of about the same age not engaged in the ancient practices.

The Emerald Tablet
Tabula Smaragdina

It is true,
that as it is above,
so it is below.
All things are from the One,
by the One becoming Two.

~

The Sun is the Father,
the Mother is the Moon.
The breath has carried it to the belly
and the body nurtures.

~

The gateway to perfection is now opened
through this potential
lying in the depths of the physical.
Only with the greatest care,
is the physical separated from the subtle
or the subtle from the physical.

~

It rises from the depths to the heights;
descends,
while the higher and lower
magnify the power.

~

The promise is as follows:
The world in all its glory
is seen with clarity and wisdom.
More powerful than strength and force,
solid and subtle are conquered,
and thus all is created,
that which is,

~

and the perfection of tomorrow.

The Scaffold for Perfection[150]

Robert L. Peck

The outer world reflects the inner.
Both come from mind. Mind is two.

~

One is masculine, the other feminine.
The masculine is the matrix; the feminine is the praxis.
Cultivation and its bondage are the forced juncture of
the two. Freedom starts with finding
the masculine and feminine.

~

The feminine is that which was recast as prolificacy.
The masculine is buried under conceit.
The two must then be carefully separated
and then renewed.

~

The masculine is activated with fervent yearning.
Yearning starts with the loss of immediacy.
The feminine is regenerated with
boundless childish joy.
With unrestrained sensual stimulation,
the future is desired.

~

Seeking more, masculine and feminine are increased.
The masculine begins the vision of heaven
The feminine becomes a voluptuary.

~

The two recombine to form the reality of heaven.
Inhabitants of heaven seek oneness.
This begins with union of vision and joy.

~

Eternal evolution and ecstasy is the gift.

---

[150] Peck et al., (2004) Chapter 12

List of Sources

Adkins-Regan, E. (2005) *Hormones and animal social behavior*. Princeton, NJ: Princeton University Press.

Aristotle. (1993) *Metaphysics*. In *The library of the future*. Garden Grove, CA: World Library, Inc.

Aristotle. (1993) *Youth and Old Age*. In *The library of the future*. Garden Grove, CA: World Library, Inc.

Aristotle. (2004) (J.A.K.Thomson, Trans.). *Aristotle: The Nicomachean ethics*. London, England: Penguin.

Berk, L., Felten, D. L., Tan, S. A., Bittman, B. B. & Westengard, J. (2001) Modulation of neuroimmune parameters during the eustress of humor-associated mirthful laughter. *Alternative Therapies*, *7*(2), 62-76

Cohen, R. (1984) *Acting one*. Mountain View, CA: Mayfield Publishing.

Cooper, J. R., Bloom, F. E. & Roth, R. H. (2003) *The biochemical basis of neuropharmacology*. New York, NY: Oxford University Press.

Crockett, M. J., Clark, L., Tabibnia, G., Lieberman, M. D. & Robbins, T. W. (2008) Serotonin modulates behavioral reactions to unfairness. *Science*, *320*(5884),1739.

Diamond, E. (2006) *The science of dreams*. Garden City, NY: Doubleday.

Digambaraji, S. (Ed.) (1970) *Hathapradipika of Svatmarama.* Maharashtra, India: K.S.M.Y.M Samiti.

Donnerer, J. & Lembeck, F. (2006) *The chemical language of the nervous system*. Basel, Switzerland: S. Karger AG

Dunbar, R.I.M. & Schultz, S. (2007). Evolution in the social brain. *Science, 317*(5843),1344-1347.

Ende, M. (1985) *Momo*. Garden City, NY: Doubleday.

Frey, W. H. & Langseth M. (1985) *Crying*. Minneapolis, MN: Winston Press.

Goble, F. G. (1970) *The third force*. New York, NY: Grossman.

Gray, H. (1901) *Gray's anatomy.* Philadelphia, PA: Running Press.

Griffith, R. T. H. (1987) *Hymns of the Ṛgveda*. Delhi, India: Munishiram Mantoharal Publishers.

Hamilton, A. J. (2008) *The scalpel and the soul*. New York, NY: Penguin.

Hart, A. D. (1995) *Adrenaline and stress.* Dallas, TX: Word Publishing.

Hermes Trismegistis. (1973) "The Emerald Tablet" in *Alchemy:The Secret Art* by Stanislaus K. DeRola. New York: Bounty Books

Hesiod. (1993) *Theogony. In The library of the future.* Garden Grove, CA: World Library, Inc.

Jahn, R. G. & Dunne, B. (1986) On the quantum mechanics of consciousness with application to anomalous phenomena. *Foundations of Physics, 16*(8),721-772

Jahn, R. G. & Dunne, B. (1989) *Margins of reality*. San Diego, CA: Harcourt Brace.

Ladas, A. K., Whipple, B., & Perry, J. D. (1982) *The G spot*. New York, NY: Dell.

Lambdin, T. O. (1988) (Trans.) The gospel of Thomas. In J. M. Robinson, (Ed.), *The Nag Hammadi library.* (pp.126-138). San Francisco, CA: Harper and Row.

Liddell, H.G. & Scott, R. (1996) *A Greek-English lexicon: Abridged from Liddell & Scott's Greek-English lexicon*. Oxford, England: Clarendon Press (Original work published 1891).

London, J. (1915) *The star rover*. New York, NY: Macmillan.

Lowrie, W. (1923) *Monuments of the early church.* New York, NY: Macmillan.

Luk, C. (1970) *Chan and Zen teaching.* Berkeley, CA: Shambala.

Lucretius. (1993) On the nature of things. In *The library of the future*. Garden Grove, CA: World Library, Inc.

Matsumoto, M. (1988) *The unspoken way.* Tokyo, Japan: Kodansha Intl.

Monier-Williams, M. (1990). *A Sanskrit-English dictionary* Oxford, England: Clarendon Press. (Original work published 1899).

Morris, W. (Ed.) (1969) *The American heritage dictionary*. New York, NY: Houghton Mifflin.

Nilsson, M. P. (1957) *Dionysiac mysteries of the Hellenistic and Roman age.* Lund, Sweden: Gleerup.

Pandit, M. P. (Ed.) (1977) *Service Letters*. Ponticherry, India: Aurobindo Ashram.

Parsons, J. D. (1896) *The non-Christian cross*. London, England: Simpkin, Marshall and Co.

Pascal, B. (1993) Pensées. In *The library of the future*. Garden Grove, CA: World Library, Inc.

Peck, R. L. (1976) *American meditation and beginning yoga.* Lebanon, CT: Personal Development Center, 1976.

——— (1985) *Handbook for goats.* Lebanon, CT: Personal Development Center.

——— (1998) *The golden triangle.* Lebanon, CT: Personal Development Center.

Peck, R. L., Cassinari, L. M. & Gavlick, C. S. (2004) *Joy and evolution*. Lebanon, CT: Personal Development Center.

——— (2006) *Directing Life*. Lebanon, CT: Personal Development Center.

Peirce, C. S. (2000) *The philosophy of Peirce*. J. Buchler (Ed.). New York, NY: Dover.

Plato. (1989) *Symposium of Plato*. (T. Griffith, Trans.) Berkeley, CA: University of California Press.

Plato. (1993) *Laws*. In *The library of the future*. Garden Grove, CA: World Library, Inc.

Plato. (1993) *Phaedrus*. In *The library of the future*. Garden Grove, CA: World Library, Inc.

Plato. (1993) *Republic*. In *The library of the future*. Garden Grove, CA: World Library, Inc.

Plotinus. (1991) *The enneads*. (S. MacKenna, Trans.). London, England: Penguin.

Prigogine, I. (1996) *The end of certainty*. New York, NY: Free Press.

Provine, R. R. (2000) *Laughter*. New York, NY: Penguin Books.

Putnam, R. D. (2000) *Bowling alone*. New York, NY: Simon and Schuster.

Sapolsky, R. (2003, Sept.)Taming stress. *Scientific American, 288*, 87-95.

Resnick, N. M., & Griffiths, D. J. (2003) Expanding treatment options for stress urinary incontinence in women. *JAMA*, *290*(3), 395-7.

Riddihough, G., & Zaun, L. (2010) What is epigenetics? *Science, 330*(6004), 611-631.

Sarasvati, P. & Vidyalankar, S. (Trans.). (1977) *Ṛgveda samhitā*, Vol. 1. New Delhi, India: Veda Pratishthana.

Schrödinger, E. *What is life?* (1992) Cambridge, MA: Cambridge University Press.

Singh, J. (Trans.) (1988) *Parātriṃśikā vivaraṇa*. Delhi, India: Motilal Banarsidass.

Solter, A. J. (1984) *The aware baby*. Goleta, CA: Shining Star Press.

Strong, J. (Ed.) (1990) *Strong's concordance*. Nashville, TN: Thomas Nelson.

Stross, B. (2007) The Mesoamerican sacrum bone: Doorway to the otherworld. *FAMSI Journal of the Ancient Americas*.

http://research.famsi.org/aztlan/papers_index.php
The Catholic encyclopedia. (n.d.) Retrieved from http://www.newadvent.org/cathen/index.html

Ulansey, D. (1989) *The origins of the Mithraic mysteries*. New York, NY: Oxford University Press.

Vertosick, F. T. (2002) *The genius within: Discovering the intelligence of every living thing.* New York, NY: Harcourt.

Wagner, U., Gais, S., Haider, H., Verleger, R., & Born, J. (2004) Sleep inspires insight. *Nature, 427*, 352-355.

Index

www.ingramcontent.com/pod-product-compliance
Lightning Source LLC
LaVergne TN
LVHW050628100826
845148LV00011B/1782